Boston Studies in the Philosophy and History of Science

Volume 285

More information about this series at http://www.springer.com/series/5710

Yuval Dolev • Michael Roubach
Editors

Cosmological and Psychological Time

Springer

Editors
Yuval Dolev
Department of Philosophy
Bar-Ilan University
Ramat Gan, Israel

Michael Roubach
Department of Philosophy
Hebrew University of Jerusalem
Jerusalem, Israel

ISSN 0068-0346 ISSN 2214-7942 (electronic)
Boston Studies in the Philosophy and History of Science
ISBN 978-3-319-22589-0 ISBN 978-3-319-22590-6 (eBook)
DOI 10.1007/978-3-319-22590-6

Library of Congress Control Number: 2015959485

Springer Cham Heidelberg New York Dordrecht London

Printed on acid-free paper

Springer International Publishing AG Switzerland is part of Springer Science+Business Media (www.springer.com)

Contents

Introduction

"There's no such thing as the time of the philosophers."[1] So Einstein concluded his public debate with Bergson in Paris in 1922. He would later add, "There remains only a psychological time that differs from the physicist's." These comments, coming from a luminary like Einstein, affirmed a bifurcation of time into "lived" or "psychological" time, and "physical" or "cosmological" time, a cleft that would only widen as the twentieth century progressed, and would determine the course of the philosophical study of time. As Canales describes it in her contribution to this volume, the clash between Einstein and Bergson was many-sided. In addition to philosophical differences, political and temperamental incompatibilities between the two spawned a personal animus. Einstein's and Bergson's successors in the debate have followed them in perpetuating not only the substantive disagreements, but also this mutual mistrust.

The debate, it must be stressed, was not about science—Bergson repeatedly emphasized that he had no objection to the mathematics or the physics of relativity theory.[2] Rather, the debate pivoted on relativity's philosophical significance, and specifically, its implications for our understanding of temporality. At the time the debate took place, a chasm was already developing between what would come to be known as the continental and analytic schools in philosophy, the former emerging primarily in France and Germany, the latter in the English-speaking world. Taking advantage of Einstein's and Bergson's celebrity status, these nascent schools trumpeted the two thinkers as paradigmatic representatives, both with regard to their substantive positions and with regard to their respective approaches to philosophical inquiry. The ongoing debate over the philosophical significance of relativity has both contributed to the shaping of each school's distinctive character and exacerbated the divide between them.

More fundamental than the debate over relativity, however, is the crucial difference in the role played by the concept of time in the two traditions. Whereas in the

[1] Einstein's actual words, uttered in French, were "Il n'y a donc pas un temps des philosophes."

[2] For a comprehensive history of the debate and its profound and far-reaching impact on culture, see Canales 2015.

analytic tradition time seems to be of secondary significance, and is considered relevant only to strictly-demarcated investigations, in the continental tradition time plays a salient role, figuring prominently in practically every field of inquiry. The notion of time is at the heart of the philosophies of Bergson, Husserl, Heidegger, Levinas, Deleuze, and others. Jack Reynolds (2009) goes so far as to suggest that the centrality of the concept of time is the continental tradition's quintessential feature.

Why is the notion of time so fundamental to the continental outlook? It is principally due, we contend, to the intimate connections between time, consciousness, and the notion of the self or subject. Consciousness of time is, of course, also discussed within the analytic tradition, but there we do not find the idea that the notion of time is inseparable from its relationship to consciousness. Within the continental tradition, time is always related, in one manner or another, to "lived time"; it is inextricable from the way it is experienced and understood by a subject or group of subjects.

In contrast to continental philosophy's preoccupation with "lived time," the analytic tradition focuses on what it refers to as "physical time." It deems the human experience of time to be marginal, if not utterly irrelevant, to "real" time and our investigation of it. Analytic philosophers seek to analyze the passage of time and tense "in themselves," that is, just as they are, independently of any mind or experience. Consider for a moment the impassioned Presentism/Eternalism controversy. It pertains chiefly, if not exclusively, to ontology. Presentists maintain that only what is present exists, and ascribe a kind of ontological superiority to present events over those that are not present. Eternalists contend that all events are on an ontological par, that is, they deny that there is any ontological difference between past, present, and future events, and reject the reality of temporal passage. This position is often identified with the "block universe" picture of time, which many take to follow more or less straightforwardly from relativity theory. On this picture, experience, being part of the world, has to be accommodated within the block universe. But it has the same status as any other natural phenomenon, and certainly plays no role in constituting time. Though Presentism differs from Eternalism in being motivated primarily by a desire to capture the phenomenology of the experience of tense and passage, like Eternalism it does not ascribe to experience any special status vis-à-vis the ontological claims it ultimately champions. It deems the present's uniqueness purely ontological, and independent of human experience.

Increasingly, however, theorists, ourselves included, have expressed concern that the analytic and continental thinkers' mutual unfamiliarity with each other's approach to time—unfamiliarity that is not infrequently accompanied by hostility—is unwarranted and counterproductive. There seems to be growing recognition that a satisfactory understanding of temporal issues cannot be achieved if either "human" or "physical" time is ignored or reduced to the other. Driven by the sense that the estrangement between the traditions has to come to an end and these two parallel lines of inquiry should be encouraged to change course and approach one another, we established a research group with representatives of both schools. The group met 20 times over the course of two years at the Van Leer Jerusalem Institute, and jointly read texts from both traditions. Its work culminated in an international conference on "Cosmological and Conscious Time."[3] This volume, which presents the proceedings of the conference, seeks to contribute to the reshaping of the philosophy of time, and to overcoming the rift that has split the exploration of time into two virtually distinct disciplines.

Each of the papers collected here attests, we believe (sometimes unbeknownst to its author), to the inextricable interweaving of physical time and human time.

We have chosen to arrange the contributions into three sections. Those in the first section address relativity theory, those in the second focus on the experience of transience, and those in the third expand on the notion of temporality within the phenomenological tradition.

Relativity Theory

Dennis Dieks's paper could, at first glance, be classed together with those in which philosophers of physics adduce the block universe picture from relativity theory. However, as Dolev suggests in his contribution to the volume, Dieks's paper, in contrast to other defenses of the block universe picture, is intensely engaged with the phenomenology of the temporal aspects of experience. Dieks's stance on relativity is refreshing in that it rejects the assumption, held not only by critics of the block universe but also by many of its proponents, that the block universe picture conflicts with our experience. Rather than accepting the Presentist's depiction of experience and acknowledging a clash between experience and theory, Dieks conducts a phenomenological study of his own, and argues that the relativity-inspired block universe picture is entirely compatible with human experience. Philosophers of physics tend to downplay, to put it mildly, the significance of "ordinary experience," and to ignore the costs of declaring experience to be mistaken or illusory. Dieks, on the other hand, takes everyday experience to be crucial for any scientific or philosophical endeavor. Far from marginalizing experience as prone to error, he argues that experience manifests that which later, when we put on our scientific hats, relativity only confirms. According to Dieks, even before relativity is taken into consideration, phenomenological analysis already shows that simultaneity plays no role in experience, and that the notion of a global "now" does not, and cannot, figure in experience. That no such notion can be accommodated within a relativistic framework thus merely reiterates what phenomenological investigation reveals independently. This is also the case with respect to the notion of time's passage. Passage is not the illusion that other defenders of the block universe too hastily claim it to be. Rather, Dieks argues, a survey of experience shows that passage amounts to nothing more than local becoming, a notion that can be made to fit comfortably in a

[3] We are greatly indebted to the Israel Science Foundation for its generous support of the conference.

relativistic setting. Here too, relativity only gives scientific credence to that which careful phenomenological analysis has already established.

Yuval Dolev's paper, while commending Dieks's attention to the manifestations of temporality in experience, takes issue with his phenomenology. Dolev claims that Dieks's attempt to construe experience as in harmony with the block universe does not succeed. The notion of "becoming," which bears all the weight in Dieks's phenomenology, does not fully capture tense and passage as we know them from experience. Dolev suggests that relativity theory became more or less synonymous with the block universe conception because its interpreters focused almost exclusively on the geometry of Minkowski spacetime, from which tense and passage are absent, while largely neglecting to investigate the nature of the present. Such investigation, he claims, reveals the present to be an integral component of the fundamental structure of reality. Yet it also establishes that the present completely lacks the character Presentists attribute to it. On Dolev's analysis, the present, in contradistinction to the Eternalists' merely *experienced* present, is part of the world, and can, contra the Presentists, be readily squared with relativity theory.[4] Moreover, Dolev shows that a global "now" is entirely compatible with relativity, and explains how to make relativistic sense of the seemingly innocent but philosophically challenging notion that something is happening on Andromeda now.

Yehiel Cohen's paper offers a third perspective on relativity theory. Cohen argues that Presentism is not only compatible with relativity, but actually entailed by it. Einstein's scientific reasoning, he asserts, was guided by a form of verificationism. Presentism, Cohen contends, follows straightforwardly from a verificationistic study of relativity. In a nutshell, Cohen rejects the standard claim that the choice of ε is a matter of convention. That this choice has no empirical consequences means there's no reason to accept that $\varepsilon = \frac{1}{2}$ in all inertial frames of reference. On the other hand, using ε-Lorentz transformations, in which ε figures as a free variable, the value of which changes from one frame of reference to another, enables us, says Cohen, to establish a notion of absolute simultaneity. Since verificationism advocates identifying the present with the real, and given the option of a relativistic construal of absolute simultaneity, an interpretation of relativity that is faithful to the verificationist spirit underlying it should endorse Presentism, understood in terms of an absolute simultaneity relation. True, this relation forever eludes our measurements, but despite this weakness, an interpretation of relativity theory based on it is superior to an interpretation that is antithetical to what we know from experience.

Jimena Canales, too, writes about relativity, but from a very different angle. Her exposition focuses on the Bergson–Einstein controversy. As already noted, the clash between the two went well beyond philosophy. But more than a clash between two celebrities of the scientific and intellectual world, it was a clash between two approaches to inquiry. For Einstein, time was "physical time." Any other temporal phenomenon was relegated to the sphere of psychology, which entailed that it lacked objective meaning. Bergson, on the other hand, insisted that while it might be

[4] Though it should be noted that there are Presentists who believe Presentism to be compatible with relativity theory.

possible, in some contexts, to separate "human" from physical time, an inquiry that turned its back on human time could not make any discoveries about time itself. According to Bergson, the meaning humans ascribe to clocks is crucial for making clocks into what they are. Clocks, taken in isolation from the meaning they have for people, are no more related to time than any other lump of matter is. Hence, though he never contested relativity theory as a scientific theory, Bergson insisted on an approach to its interpretation that was very different from Einstein's, an approach that put "human time" at the center of the inquiry. According to Canales, the debate between Bergson and Einstein and their respective followers, far from exemplifying the Habermasian ideal of a "domination-free communication community," soon evolved into an exchange in which some were doing the talking while others were being silenced. She concludes by pondering, somewhat pessimistically, whether things have changed appreciably since then.

Transience and Experience

Barry Dainton's paper constitutes a paradigmatic example of the interface between the two disparate traditions. Specifically, it illustrates how cosmological and phenomenological considerations bear on and constrain each other. A phenomenological analysis of the experience of motion and change reveals a dynamic element that is irreducible and ineradicable. Starting in the mid-nineteenth century, an array of theories purporting to explain the structure of the experience of motion and change has been suggested. In his paper, Dainton reaffirms Extensionalism, the position that is now widely associated with his name. Extensionalism maintains that the experience of change takes place in a temporally-extended specious present and adjacent specious presents are joined by the overlapping of shared content. Dainton defends Extensionalism against objections to the notion of a "specious present" that have been raised by Dolev, and offers further arguments to establish Extensionalism's advantages over Retentionalism, one of its main rivals. Dainton then turns to the paper's core question: what cosmological conception of time fits best with Extensionalism? The block universe picture is compatible with Extensionalism, but unsatisfactory when considered phenomenologically. Presentism in its standard form, namely, the view that the world consists of a series of point-like presents, cannot be squared with Extensionalism: the volumeless present cannot sustain an extended specious present. Dainton then points out a variation on Presentism, namely, Extended Presentism, which provides a hospitable environment for Extensionalism. In view of recent developments that seem to be reintroducing a real present into cosmology, Dainton suggests that a form of Extended Presentism may be the most promising option for a cosmology that is both supported by empirical research and informed by a sound phenomenology of experience.

Jenann Ismael argues that our experience of time does not necessarily conflict with time as interpreted by post-relativistic physics. We experience time as flowing, passing, and manifesting asymmetry between past and future. Physical time, on the other hand, lacks these features: the universe is conceived as a block. Ismael's strategy is to show that these seemingly, opposed points of view do not necessarily conflict. She draws on an analogy between time and space. Viewing time from a particular moment is like viewing space from a particular location. There is no conflict between this experience of space and absolute space. Concerning time, Ismael argues that, viewed at a particular moment by an embedded, embodied participant, events are ordered by their practical and epistemic relations to the viewer at different points in her life. Strung together in temporal sequence, they produce a changing image of a world with a fixed past—history—and open future, a world in the process of coming into being. Passage, flow, and openness arise as artifacts of changes in perspective, relative to the fixed backdrop of history. On the physical view of time, from an Eternalist point of view, those same events are represented in a manner that is invariant under transformations between temporal perspectives.

Tamar Levanon examines the basic tension between the unity of temporal experience, on the one hand, and the internal variation associated with the succession of moments, on the other. In particular, she focuses on the role of our raw sense of flow and transition within the conceptualization of temporal experience. In seeking to understand this sense of flow, she examines the models put forward by William James and Alfred North Whitehead, both of whom accept a basic experience of transition, and contrasts them with Bertrand Russell's approach. She then discusses James's and Whitehead's respective proposals for resolving the tension between unity and multiplicity in temporal experience. James and Whitehead share the view that time is made up of extended yet atomic units of duration, and both try to construct an account of temporal continuity that is consistent with our experience of temporal units. But there are also serious disagreements between them. One concerns the internal structure of each drop of experience, to use Whitehead's term. Another pertains to the nature of relations and specifically to the character of the relations between successive drops. These disagreements entail different descriptions of temporal continuity, which we can invoke to revisit the gap between analysis and experience.

The proposal that presentness can be fleshed out in terms of consciousness is tendered for consideration in Ulrich Meyer's paper. But his aim is far broader—he seeks to establish that no characterization of presentness will be satisfactory. In particular, no phenomenological study of time can yield a tenable notion of presentness. Time, he contends, is nothing over and above physical time, about which we should be Eternalists. Meyer's argument hinges on the simple and quite uncontroversial observation that for any instant of time t, it is the case that at t, t is present. Meyer contends that given this innocuous truism, it is impossible to characterize a particular moment as present in a way that would distinguish it from any other moment. Establishing this conclusion, however, requires the rebuttal of theories that, in attempting to articulate the contrast between the present moment and moments that were or will be present, use either tensed logic or classic tenseless logic.

In their paper, Meir Hemmo and Orly Shenker argue that temporal directionality cannot be derived from science, but must be added to it. Their thesis emerges from the ongoing vibrant discussion of the time-reversal invariance of the laws of nature. But perhaps more than other participants in the debate, Hemmo and Shenker take human time very seriously. Indeed, their argument pivots on the temporal directionality of human experience, the undeniable fact that "there is a clear distinction between past and future in our experience." Starting out from experienced directionality, they go on to argue that temporal directionality is a fundamental feature of the structure of physical reality. As long as science fails to account for this feature, it is not complete. But the laws of nature, being time-reversal invariant, cannot be invoked to account for temporal directionality. Hence, physics must be externally supplemented with temporal direction. Without such augmentation, no interpretation of a scientific theory, no derivation from science of claims about the world, can be satisfactory. En route to this conclusion, Hemmo and Shenker show that the Past Hypothesis (which says that the universe's initial state was characterized by low entropy), often appealed to in attempts to engender temporal direction from within physics, in fact presupposes temporal directionality.

Temporality and Phenomenology

Michael Roubach's article discusses Heidegger's notion of "primordial temporality" and its relation to other notions of time, focusing on Heidegger's claim that primordial temporality is the basic notion of time, and that other notions of time, and in particular, what Heidegger calls the "ordinary" (*vulgäre*) notion of time, presuppose it. Paul Ricoeur, and more recently, William Blattner, have asserted that Heidegger does not provide a satisfactory argument in defense of this claim. Roubach disputes this claim, and shows that it is, in fact, possible to construct a valid argument for the primacy of primordial temporality. Roubach's suggested argument adduces assumptions that can be attributed to Heidegger on the basis of his writings, though Heidegger did not articulate them explicitly. These assumptions pertain to the arithmetization of the continuum and concepts related to that arithmetization, in particular, Brouwer's intuitionistic notion of choice sequences. The paper concludes that Heidegger's notion of primordial temporality, which differs from both conscious time and cosmological time, can play a useful role in efforts to bridge the dichotomy between the latter two conceptions of time.

The discussions about time within what is usually called the "continental tradition" do not constitute a unified field. But in some cases the similarities within this tradition are greater than they may initially seem to be. Philip Turetzky's paper points out several affinities between the Deleuzian and Husserlian concepts of time. Moreover, Turetzky asserts, comparing the views of Deleuze and Husserl can enhance our understanding of Deleuze's approach to time. Turetzky's main focus is Deleuze's account of the three syntheses of time. According to Deleuze, three passive syntheses generate and constitute time. Each synthesis centers on one tense, though it also embraces the others. The basic tense of the first synthesis is the present, of the second synthesis, the past, and of the third synthesis, the future. In presenting these syntheses ordinally and regressively, that is, as proceeding from the "conditioned" to its "condition" (to use Kantian terminology), so that the second synthesis is the condition for the first, and the third synthesis the condition for the first and second, Turetzky upholds the prevailing view. But he adds that with regard to the order of genesis, the third synthesis comes first. Turetzky argues that the second synthesis of time bears a close affinity to Husserl's notion of retention, and he adduces Husserl's conception of time-consciousness to argue that in the order of genesis, Deleuze's first synthesis depends on the second. The main difference between Deleuze and Husserl arises with respect to the third synthesis. Turetzky shows why the third synthesis is required for the second synthesis, and therefore for Husserl's conception of time as well.

Dror Yinon's paper offers a different perspective on Deleuze's conception of time. Yinon begins by analyzing the framework of the transcendental and phenomenological traditions. These traditions purported to show, first, that objective time is grounded in subjectivity, hence subjectivity should account for objective time; and second, that subjectivity is intimately related to the transcendental structure of temporality. Yinon argues that while Deleuze indeed espouses the basic transcendental stance that time is inseparable from subjectivity, he goes beyond that, arguing that inquiry into the relations between subjectivity and time reveals that different layers of subjectivity are generated by three different syntheses of time. Yinon then discusses the three syntheses from the perspective of time's role in constituting the kind of subjectivity that emerges; differing temporalities make for different layers of subjectivity. Yinon pays special attention to Deleuze's notion of change. Deleuzian change, he contends, does not ensue from events that befall already-constituted entities, but rather functions as a transcendental productive principle of any event that might take place. The last part of Yinon's paper examines Deleuze's idea of change by invoking McTaggart's notorious critique of time as change. This move illustrates the benefit that can be gained by the mutual critical assessment of ideas from both the analytic and continental traditions, at least with regard to questions of time. We believe that similar benefits will accrue from such "cross-traditional" assessment of other questions as well.

Bar-Ilan University, Ramat Gan, Israel Y. Dolev
Hebrew University, Jerusalem, Israel M. Roubach

References

Canales, J. 2015. *The physicist and the philosopher: Einstein, Bergson, and the debate that changed our understanding of time*. Princeton: Princeton University Press.

Reynolds, J. 2009. Continental philosophy and chickening out: A reply to Simon Glendenning. *International Journal of Philosophical Studies* 17: 255–272.

Part I
Relativity Theory

Chapter 1
Physical Time and Experienced Time

Dennis Dieks

Abstract In our direct experience time is strikingly different from space: time has a dynamical aspect that space completely lacks. This feeling of flow and passage is well represented in the A-theory of time, in which the concept of a moving Now is central. Given the clarity and immediacy of our experience of temporal flow it seems that the rival "static" B-theory, in which there are only unchanging temporal relations, starts with an enormous handicap: whereas the A-theory directly explains our experience of time, the B-theory must apparently squirm itself out of the problem of explaining it away as an illusion.

However, on second thought it is not so clear how the A-explanation of our experience is supposed to work. Even if the A-theory were to be correct and the flow of time an objective feature of reality, there still is the question of how this objective motion of the Now could make itself felt in our apperception of time. The problem is that the concepts of objective passage and becoming that are central in the A-theory do not make contact with anything we know about how natural processes work and therefore cannot help us to understand our perception in a naturalistic, scientific way. This problem is acutely illustrated when we look at recent philosophical work on the A-theory. These approaches usually accept tenses as a primitive concept (by employing primitive tense operators). But these primitive tenses do not relate to what we know about how time perception functions; from a physical or physiological point of view it is mysterious how primitive tenses could help to explain our intuitions.

By contrast, the concepts used in the B-theory of time do connect with scientific theory. Perhaps surprisingly, explaining our experience of passage has better prospects in the B-theory than in the A-theory.

Keywords Experienced time • Explanation of time experience • B-theory of time • Passage • Time flow • Time in physics

---

D. Dieks (✉)
History and Philosophy of Science, Utrecht University,
PO Box 85.170, Utrecht 3508 AD, The Netherlands
e-mail: d.dieks@uu.nl

Y. Dolev, M. Roubach (eds.), *Cosmological and Psychological Time*, Boston Studies in the Philosophy and History of Science 285,
DOI 10.1007/978-3-319-22590-6_1

1.1 Introduction

In our direct experience there is a striking difference between time and space: we perceive time as dynamic whereas space is static. McTaggart famously represented this dynamic aspect of time in his "A-series". According to the A-theory of time, time's passage is characteristic of time itself, independently of any experience of it: time in itself really flows, with a Now that moves from Past to Future.

By contrast, McTaggart's B-series only recognizes temporal *relations*, such as "earlier than" and "later than" and invokes neither a privileged Now nor a notion of flow (McTaggart 1908; Oaklander 2004; Dolev 2007; Dainton 2010). In the B-theory of time, the temporal dimension is represented in the same way as the spatial dimensions: events stand in the earlier-later relation like they stand in the relation of taking place to the left or right of each other. This B-theory accords with how time occurs in physical theory. In the equations of mathematical physics, both classical and relativistic, temporal coordinates occur in the same way as spatial coordinates, and at no point is there an appeal to an additional *dynamic* aspect of time. In accordance with this, there is no privileged Now on the physical time axis, just as there is no Here in space.

The A-theory does operate with a privileged Now: this Now shifts from Past to Future. This A-picture of time appears to account fully for our direct experience of passage—whereas the B-theory seems to be deficient on this score. The explanation offered by the A-theory is simply that our experience is faithful to reality itself, time actually passes precisely as we experience it. So at first sight there seems to be a strong argument here for the priority of the A-theory: the A-theory is both descriptively and explanatorily more complete than the B-theory (cf. Zimmerman 2008).

It is the purpose of this chapter to critically analyze the just-sketched argument in favor of the A-theory. First we shall review in some detail the way physics deals with time, both in classical mechanics and in relativity theory, in order to get a grip on the explanatory resources of physics (and fundamental science in general) *cum* B-theory with respect to our temporal experience, in particular our experience of passage. An important observation will be that the accounts given by physics of processes in space and time are *local*, which excludes a direct explanatory role for a global notion of now or, in other words, global simultaneity. This may seem to confirm the explanatory superiority of the A-theory, because we possess a strong intuition to live in, and be in contact with, an extended Now. However, we shall argue that this superiority of the A-theory and the concomitant explanatory inadequacy of the B-theory are illusory: there is no reason to think that the B-theory is unable to explain the intuitions we have about the Now. In fact, there is an uncontroversial consensus that our intuition of being in causal contact with a global Now is non-veridical—this clear example of non-correctness of A-type intuitions should warn us that more may be in store.

In fact, A-type notions would be superfluous if the B-theory were able to accommodate and explain what we actually experience; it would in this case not be an objection but rather a success if the B-theory succeeded in accomplishing its explanatory task without positing the presence of a flow of time in physical reality itself. As we shall argue, there is no reason of principle to believe the B-theory cannot achieve this success. By contrast, it is less than clear how the A-theory can redeem its promises of explanation here. Even if we assumed that the A-picture of time makes perfect sense and reflects what the world is like, namely equipped with a moving Now, it would remain obscure how this could play a role in the explanation of our experience of passage.

1.2 Time in Newtonian Physics

Both in Newtonian physics and in special relativity events are placed in a pre-given spatiotemporal arena, in which the position of each space-time point can be fully fixed by four coordinates. Also both in Newtonian and special relativistic physics, this space-time background against which all physical processes are described possesses spatiotemporal properties by itself: it has a well-defined geometrical structure. Focusing on Newtonian space-time first, we can easily understand how this spatiotemporal background figures in explanations. Indeed, the laws of classical mechanics would not make sense without a definite space-time structure. Consider the law of inertia: a body on which no forces are exerted moves uniformly in a straight line or remains at rest. In order that this statement possesses content, it must make sense to distinguish between straight lines and curves, and for this we need the concept of spatial *distance* (a straight line is the shortest connection between points, so with the help of distances curves and straight lines can be distinguished from each other). More important for our subject here, to give meaning to "uniform motion" it must not only be clear what equal distances are, but a definition of equal periods of *time* must be available as well. In order to make sense of Newtonian mechanics we need a definition for the congruence of temporal intervals.

The resulting Newtonian space-time can be pictured as a stack of three-dimensional spaces-at-a-time, all being copies of each other and each equipped with Euclidean spatial geometry. Between these instantaneous spaces there exist well-defined time intervals, and also a mapping that defines the identity of spatial points through time. As a consequence, between any two events in Newtonian space-time there is both a definite spatial and a definite temporal distance.

Newton famously commented on time in the just-defined spatiotemporal framework that it "flows equably". But we should note that only the "equably" in this statement may to some extent be taken to actually reflect a feature of the structure of Newtonian space-time, namely its invariance under temporal translations (in fact, Newtonian space-time is completely homogeneous, and therefore also invariant under translations in space). By contrast, a "flow" of time is not defined within the mathematical framework that we have just described. In order to define such a *flow of time* we need two things: first, something that is doing the flowing, something that executes the flow-motion; and second, we need something with respect to which the change in the flowing subject can be defined. Both things are missing in Newtonian space-time.

Clearly, what is *intended* as the thing that flows is the Now. It is true that there are global nows (with small letter n!) in Newtonian space-time: in fact, each space-at-a-time instantiates such a now. But there is an uncountable infinity of such nows (just as there are uncountably many points on the spatial axes or real numbers on the real line). Newtonian space-time is the infinite stack of these nows, there is nothing in the Newtonian spatiotemporal structure over and above them. In particular, there is no *privileged* Now that could "go" from one now to another, while keeping its identity.

Moreover, and this relates to the second missing element, if we nevertheless try to imagine a hypothetical special Now, it is clear that it should "visit" each and every now; the Now is supposed to flow past all nows. But this can only mean, within the Newtonian picture as defined above, that all nows are equally entitled to be considered as the Now. It does not make sense to say that the Now can shift from one now to another. It can only be at any now at precisely the instant defined by that individual now; it would be self-contradictory to assume that the Now can find itself at a certain now at some other instant than defined by that particular now. But this in turn means that the Now has to be located at each now; the Now should coincide with each and every now!

Another way of making the same McTaggart-like point is that in Newtonian space-time there is no "supertime", an additional independent time variable, with respect to which a changing position of the Now among the nows could be defined. Therefore, the very definition of Newtonian space-time entails the inapplicability of the notion of the "flow of the Now".

The introduction of a supertime T would legitimize statements like "At supertime T the Now finds itself at the now $t = T$", but without a role for T in scientific theories this would not enhance our possibilities of scientific explanation. However, from the side of science complaints about explanatory failures because of the absence of supertime have never been heard and this raises the question of whether it makes sense to introduce this new T. Finally, the introduction of such a supertime would not really inject the dynamic aspect the A-theorist desires: what we are actually doing is adding one additional dimension, and in the new five-dimensional continuum the same questions about flow can be asked as before. We can only end up in an infinite regress when we attempt to do justice to the notion of passage by adding new temporal dimensions.

1.3 Relativity Theory

In physics Newtonian space-time has of course been superseded and we should investigate the conceptual resources of relativity theory. Let us first focus on special relativity. Like pre-relativistic physics this theory positions all events in the history of the universe within a spatiotemporal continuum, a four-dimensional manifold of space-time points. As in the case of Newtonian space-time, this spatiotemporal continuum of special relativity—Minkowski space-time—possesses a definite geometrical structure that is given *a priori*, in the sense that it is independent of the material content of space-time. This new geometrical structure derives from a distance function, **ds**, between neighboring space-time points. The most important difference with Newtonian space-time is that this distance function is no longer a spatial distance *plus* a temporal distance: instead we now have just one number for the length of the space-time interval between space-time points. It is true that this four-dimensional distance can be decomposed into spatial and temporal parts, but this decomposition is not given by Minkowski space-time itself—we need the specification of something external, a system of coordinates, or a frame of reference, in order to split up **ds** in definite spatial and temporal distances. Minkowski space-time by itself does not define any preferred coordinate systems, and this is behind the statement that in special relativity the distance and time intervals between any two given events are "relative". If the worldline is added of someone who travels between two events, the spatial distance between the events as judged from this worldline is zero. In this case **ds** along the worldline represents the time interval between the events, as measured by the traveler. For another observer who is moving with respect to the first one but also considers himself at rest, the spatial distance between the same events will not vanish, and for this second observer the time difference between the events will be less than **ds**.

In Minkowski space-time in itself, without the addition of a point of view or some specific curve connecting the events, there is only the spatiotemporal distance function **ds** between neighboring space-time points, without a splitting into a spatial and a temporal component. This pre-given four-dimensional distance function **ds** suffices to do all physics in relativity theory.

A corollary of the absence of a preferred time versus space splitting is that there exists no unique *time function* on Minkowski space-time, i.e. a function that assigns a unique and physically meaningful time value to each space-time point. This is quite different from the situation in classical Newtonian physics. According to Newtonian theory, once we have chosen any particular event as our time origin, and have decided on a time unit, each event in the history of the universe can be assigned a definite time; this defines a time function on all space-time points. The physical interpretation of this time function in Newtonian theory is simple: the time difference between any two points in space-time is the time taken by any journey between the points in question, i.e. the amount of time by which the traveler has aged. But according to relativity theory we cannot consistently assign one physically meaningful time value to events in this manner. Relativity theory operates with the integral of the distance function, $\int$**ds**, between pairs of points on a curve that represents a journey (a world line) between the events to represent the time taken by the journey. For any pair of points on a world line, $\int$**ds**, calculated between these points along the world line, has the physical interpretation of the lapse of time that would be measured by a clock whose motion between the two point events is represented by the worldline in question.

Time thus enters special relativity in the form of the *duration* of physical processes, and needs the specification of such a process (along a world line) to become definite. It is this time that occurs in the special relativistic equations and governs the evolution of physical processes.

It is essential for relativity theory that the time interval between two events, defined in this way, depends on the path in Minkowski space-time that connects the two events: different paths are associated with different time intervals. The resulting non-uniqueness of the physical time difference between events is what is behind the non-existence of a time function on the points of Minkowski space-time. In Minkowski space-time *per se*, without the addition of external structure (worldliness, processes that take place) there *is* no "*time* interval" between any two space-time points.

A concrete illustration of this basic feature of relativistic time is the notorious case of the twins, one of whom stays on Earth while the other departs on a space journey and eventually returns: there is no fixed amount of time between the events of the departure and the reunion of the two twins. The space traveler turns out to return younger than his or her twin brother or sister, and the age difference can be made arbitrarily great by varying the path of the traveling twin.

This peculiar structure of relativistic time has an immediate consequence for the notion of *simultaneity* in relativity theory. Suppose we take some event as our origin in Minkowski space-time, and ask for all events that occur one unit of time later. In Newtonian space-time the answer to this question is given by the three-dimensional space-at-an-instant that is one time unit later than the original event. The points in this space realize a natural simultaneity relation (which is why we can speak about a space at an instant). For example, these points cannot be mutually connected by a signal with a finite speed (signals with finite velocities take time, which makes the point of arrival *later* than the point of departure) but can only interact via *infinitely fast* processes (that do not need time for their propagation). In relativity theory the situation is very different, as we can see by considering the twin case again. By traveling fast enough (with a speed arbitrarily close to the speed of light) along a non-inertial path we can push the events at which one time unit has lapsed arbitrarily far into the future of events that are one time unit later from the origin as measured along an inertial world line. Therefore, the collection of events "one time unit later than a given event" does not define a sensible notion of simultaneity—there exist causal signals with finite speeds between these events, establishing earlier-later relations.

Elaborating on this point, one can easily see that any notion of simultaneity that might be proposed within relativity theory cannot possess a causal significance for dynamical processes. Any acceptable simultaneity relation should satisfy the requirement that simultaneous events cannot be connected by a causal signal (such signals cannot have infinite speeds according to relativity: the maximum signal speed is the speed of light—causally connectible events stand therefore in the "earlier-later" relation to each other). But this means that by definition in relativity theory simultaneous events are unable to "feel" each other: no physical contact is possible between them. In other words, any notion of simultaneity one could think of could only group together events that are causally cut off from each other. These events consequently cannot work together and will not function as a physically coherent whole (more precisely, they cannot do so by virtue of being simultaneous; it could be, of course, that there are relations between them because of common causes in the past).

This has a consequence for experienced time: it is impossible, according to relativity theory, that we experience events that are simultaneous, according to any consistent notion of simultaneity, with the local act of observation. A global Now can therefore play no direct role in experience—assuming that our experience can be explained naturalistically and respects the laws of physics (Dieks 2006).

The more general background of what was just said is the *locality* of all physical interactions in relativistic physics. According to relativity, material bodies and fields can only feel and influence each other directly per space-time point at which they are co-present. How one groups distant events together under the denominator "simultaneous" is therefore immaterial for what happens in physical processes. In accordance with this, simultaneity does not play any role in the mathematically formulated laws of relativistic physics.

When we now compare and contrast Newtonian and Minkowski space-time, we see that the latter is even less friendly to the concept of "flow of time" than the former. In the case of Minkowski space-time the Newtonian uniquely defined spaces-at-an-instant are no longer there, so that Minkowski space-time cannot be conceived as a stack of such spaces. Related to this, the notion of "identity of spatial points over time", which was well defined in Newtonian space-time, has no place in Minkowski space-time. So there no longer is a family of global nows, a preferred "foliation", that could, at least intuitively and at first sight, be associated with the doctrine of the flow of time. Time has become a local and path-dependent quantity, so that any flow that could be made compatible with special relativity necessarily must become local and path-dependent itself. Since there are no preferred worldlines or paths defined in empty Minkowski space-time, the project of defining a flow of time on the basis of the properties inherent in space-time itself is even more hopeless than in Newtonian theory.

One can only hope to make a connection to time as we experience it by introducing specific worldlines (in particular of observers experiencing time) into Minkowski space-time and by investigating the properties of **ds** along these worldlines. Relativistic time is dependent on the presence of such material processes and is not well-defined in empty Minkowski space-time.

Even after the introduction of worldlines, along which time intervals are well-defined, the problem of the absence of a "supertime" remains exactly what it was in the Newtonian context. Any worldline just *is* the collection of the points lying on it, and any such point is equally entitled to the status of being the privileged Now. There is no independent parameter with respect to which a shift of the Now along a worldline could be defined.

Summing up, in comparison to Newtonian space-time special relativity theory introduces additional complications for the flow of time doctrine. First, everything becomes local and there is no longer a family of preferred global nows. Second, the notion of a time lapse between events only becomes definite after the introduction of a curve, a worldline that represents a material process that connects these events. Therefore, definite lapses of time are no longer inherent in the spatiotemporal framework itself, but depend for their definition on the specification of a physical process.

Finally, the inconsistency of the Newtonian notion of a flow of time is still with us: it returns here as an inconsistency in the idea of a local flow of the Now along a world line. On the one hand the existence of a flow implies that there is one privileged Now on a world line; on the other hand this Now should be present at any event exactly when it actually occurs. Since a worldline is nothing but the collection of the actually occurring events forming it, we have a contradiction here. The Now should single out one point on the world line, but it also has to be omnipresent on it—a return of McTaggart.

Special relativity is not the latest word in space-time physics, and we should at least glance at the *general* theory of relativity. We can be relatively brief about this, however, at least for the purposes of our present discussion. In general relativity there is no *a priori* fixed spatiotemporal background geometry, like there is in Newtonian physics and special relativity. The geometrical structure becomes dynamic in general relativity: the geometrical properties are determined by a "metrical field", a physical field that interacts with the material content of the universe. There are no well-defined spatiotemporal intervals in the ("bare") manifold of space-time points by itself, without this metrical field. So, even more than in special relativity the consideration of actual physical processes becomes essential to get a grip on the properties of time. There is no time *per se* according to general relativity, independently of the material content of the universe.

The global structure of space-time, once the material content of the universe has been taken into consideration (by solving the "Einstein equations"), may differ very much from that of Minkowski space-time. Intuitively, the space-time structure of general relativity stands to the space-time structure of special relativity as the geometry of a curved two-dimensional surface (perhaps with a non-standard topology, for example with holes in it) to the Euclidean geometry of the plane. One may think of a general relativistic space-time as being formed from Minkowski space-time by cutting away pieces of it, and applying deformations (changes in **ds** between points) that introduce curvature. Equivalently, one can think of general relativistic space-times as constructed from tiny pieces of Minkowski space-time sown together (in the same way that a curved two-dimensional surface can be composed of minute pieces cut from a plane). It follows that in small regions of general relativistic space-time the relations between events remain as those in Minkowski space-time, analogous to the validity of Euclidean relations in small portions of curved two-dimensional surfaces. Our earlier conclusions concerning time in relativity therefore also apply to the local situation in general relativity. Because physical laws in general relativity are no less local than in special relativity, it is sufficient to consider small portions of space-time to study the nature of causal relations, and we find that these exhibit the same structure as in special relativity. Time still enters as the duration $\int$**ds** along worldlines, and simultaneity still plays no role at all in determining the outcome of physical processes. Moreover, just as in Newtonian theory and special relativity, there is no supertime parameter that could help to make sense of the notion of flow of time. So our conclusions about time in special relativity are only reinforced when we take general relativity into account.

The above considerations did not make reference to physical (a)symmetries or the time invariance of the laws of physics (or their possible time dependence). This is with good reason: such (a)symmetries are not relevant to the topic of the flow of time in the sense of the A-theory. If physical laws were to be found that are time and position dependent, or if the material content of the universe were to show asymmetries along a time axis, this would not create room for the kind of Now that is needed for the viability of the notion of flow. A time dependence of physical laws, or a time dependence of material properties of the universe, can be introduced without any commitment to a Now that shifts from Past to Future. What such a variation in time merely establishes is a correlation between physical quantities and a time coordinate. At best, such a correlation could be used to identify an instant in terms of the instantiation of a physical state of affairs; for example, "instant 1 is the unique moment at which the entropy has value S_1". But such correlations do not help us to define a *flowing* Now: such a Now has to flow through *all* thus defined instants, so that the possibly unique characterization of now-instants by physical properties cannot single out a unique Now.

Both the A and the B theories are compatible with time independent as well as time dependent laws, and with both symmetric and asymmetric universes—the issue of time invariance and time symmetry is logically independent of the question of whether or not there exists a moving Now.

This is not to deny that time asymmetry can play an important role in the physical explanation of the difference between how we experience past and future. We remember the past and do not know the future; this is a time asymmetry in itself which should be relatable to a physical time asymmetry (plausibly connected to the second law of thermodynamics). But in this article we shall focus on time *flow*, and not on the asymmetry between the two time directions (this asymmetry can be discussed completely within the B-theory of time).

1.4 Time Without Now and Explanation

An essential question that we have to face is whether physics without Now and without flow of time can be *explanatorily complete*. If the answer were to be negative, this would constitute an important argument after all for the necessity of the A-theory and the reality of the Now and its passage.

In order to investigate this question we need to be explicit about what exactly needs to be explained. Clearly, if physics is to be an explanatorily complete discipline, it needs to account for all physical properties that are instantiated during the history of the universe. This includes the details of the states of physical systems at all times and positions, and the relations between them. But we shall take a naturalistic position also with respect to observers and their observations, and require that the physical explanation shall extend (at least in principle) also to the past, present and future as we *observe* them. Most importantly for our purposes here, scientific explanations should also cover our awareness of the flow of time. If this could be accomplished, we would be in a position to explain our daily intuitions on the basis of a scientific world picture that does not take these intuitions as literally veridical. Given the internal tensions in the notion of the flow of time, when taken literally as a feature of the physical world, this would be a welcome result.

The actual explaining here is evidently an immense task, and in the final analysis probably an impossible one given the specific theories we currently possess—these theories will undoubtedly prove wrong in their details, just as Newtonian theory has been demonstrated wrong by relativity and quantum theory. But we can still investigate the possibilities of principle here. We can look at what kind of phenomena theories of the general form shared by Newtonian mechanics, special relativity or general relativity, which all work with a B-theory of time, can explain and ask whether the general nature of these explanations may be satisfactory.

The general picture yielded by these theories is as follows. The laws of physics, together with boundary and initial conditions (which do not need an appeal to a Now!) lead to exactly the kind of four-dimensional representation without flow and without Now that we have already sketched. The laws establish relations between data on surfaces of initial conditions ("Cauchy surfaces") and physical conditions elsewhere in the four-dimensional universe, regardless of how these surfaces of initial conditions are situated with respect to any supposed Now (a Cauchy surface is a three-dimensional space on which initial/boundary conditions, together with the physical laws, fix the physical state of the whole four-dimensional universe). In Newtonian physics the laws governing physical processes are such that they determine that physical processes will measure the time interval between hyperplanes of absolute Newtonian simultaneity; in special and general relativity the laws are purely local and time intervals along worldlines become the temporal determinants governing physical processes, as we have already seen. It is true that this leads to a difference in global temporal structure: in Newtonian physics events are totally ordered in time, whereas in relativistic physics the temporal order is only partial. But in both cases all events are depicted as occurring at their own space-time position in the four-dimensional manifold, characterized by their individual values of physical quantities, and without assigning a special status to a Now. In other words, we have a "block universe" representation of the history of the universe.

There can certainly be becoming and change in this picture: different events along a worldline may well and generally will display different physical properties. For example, the motion of a particle is represented by a curve in space-time (the worldline of the particle), and in general such a worldline will be characterized by different values of physical quantities along it (different values of the particle's velocity along the worldline, for instance). So there can certainly be change *within* the thus-described world. This is change in the unproblematic sense of variation of properties as a function of time—the kind of change science is about. By contrast, it does not make sense to say that the world as a whole (a four-dimensional entity) or "time itself" change. Time, needed to define change, is internal to the universe; there is no external temporal parameter that could be used to define a change of the universe as a whole.

The central question now is whether everything that we know from experience may be accommodated in this four-dimensional B-picture. It is true that there is no problem in thinking of all ordinary physical as being thus represented: The B-picture comprises all earlier-later relations, all causal links, and all processes of change and becoming in the sense that it specifies what properties are instantiated at each stage of every process. But could this "static" representation of change, in which there is no moving Now, also be sufficient to explain our direct *awareness* of change?

1.5 Physics and Experiential Time

By virtue of the locality of the laws of relativistic physics there are no direct causal connections between what happens in a point of space-time and what goes on at space-like separation from this point (i.e. in points that cannot be reached by a causal signal). That means that for purposes of causal explanation extended nows, i.e. hyperplanes of simultaneity, are irrelevant. Events at any space-time point should be explained by an appeal to events at that point itself or by signals reaching the point (with a speed at most equal to the speed of light). Signals that reach a space-time point with finite speed necessarily come from the past, not from distant events Now. The same remark applies to the physical explanation of events in a finite region of space-time: only the past of a region is explanatorily relevant. A notion of time passage that makes contact with physics should therefore not make use of distant simultaneity or a global Now. If a physically respectable notion of time flow is to have a chance of being viable at all, it should be *local* (Dieks 2006).

It follows that when we attempt to find a naturalistic explanation of human time experience, simultaneity and an extended Now are irrelevant. Any physical account of how we "feel" time should make use of our local situation in space-time and the signals from the past that influence us. It is consequently not important at all for the purposes of such a physical explanation whether or not a global foliation of space-time (in terms of spaces-at-an-instant) is possible and whether or not there is a unique preferred foliation. From the perspective of physics such foliations are irrelevant for the explanation of our temporal experiences. In particular, the existence of a global now (or Now!) must be irrelevant for the scientific explanation of our intuition that we are living in an extended and shared Now.

This conclusion implies that the proposals in the literature to introduce by hand preferred simultaneity hyperplanes in Minkowski space-time (resulting in neo-Lorentzian space-time), with the aim of accommodating our intuitions about the global passage of time, miss the mark. Such attempts are usually criticized for the arbitrariness that they involve—indeed, Minkowski space-time itself does not single out any such hyperplane (cf., e.g., Balashov and Janssen 2003). But this line of criticism does not go to the heart of the matter. The crux is that to the extent that such proposals are motivated by the desire to make contact with experienced time and our temporal intuitions, they are irrelevant non-starters because they do not *explain* anything about our intuitions.

The situation in general relativity is basically the same. Certain general relativistic universes (cosmological models, i.e. solutions of the Einstein equations) can be "foliated"; others cannot. That is, certain solutions can without contradiction be seen as stacks of three-dimensional space and others cannot. Of the universes that can be foliated certain types possess foliations that may be called preferred in view of their simplicity and symmetry (this is the case, for example, in Robertson-Walker cosmological models). It is frequently suggested that the possible existence of such foliations is directly relevant to the status of time in the corresponding universes, but from the above it should be clear that this is incorrect as far as the physical explanation of experienced time is concerned. For experienced time the causal past is the only explanatorily relevant region of space-time, both in general and special relativity.

Even within the conceptual framework of Newtonian space-time the significance of distant simultaneity for experienced time is doubtful at best. It is true that action-at-a-distance exists according to Newtonian physics, and that causal influences can therefore propagate along hyperplanes of absolute simultaneity. But in Newton's mechanics it is only gravitation that propagates infinitely fast in this way; and it is hardly plausible that our time awareness has much to do with gravitational interactions between us and the rest of the world. If we were to suppose that in the Newtonian picture also superluminal electric signals are possible (Coulomb's law), this would not help either: positive and negative charges shield each other off, so that in practice no effects of long-range Coulomb forces could be felt. So even within the Newtonian framework the explanation of our intuition that there is an extended Now with which we are in immediate contact cannot be grounded in the existence of a relation of distant simultaneity. The idea that our intuition of living in a global present provides evidence for the physical existence of a global Now rests on a misunderstanding.

This point is reinforced when we look into how physics actually explains our Now-intuition. For this, it is important to realize that the immediate contact we seem to have with spatially distant regions is in fact mediated by light: we *see* the extended Now (Butterfield 1984). Our strong feeling of immediacy is due to the fact that the speed of light is huge compared to other speeds we encounter in daily life and to the circumstance that objects around us hardly change during the temporal intervals that light needs to reach us. Everyone knows, in an abstract and theoretical way, that the speed of light is finite—but this speed is so enormous compared to the speeds of ordinary processes surrounding us that this undeniable scientific fact has not become internalized and has remained foreign to our intuition. In ordinary circumstances we cannot easily obtain information showing us that the things that we see are past. Signals coming from distant objects reach us from longer ago than signals from objects that are nearby, but under ordinary conditions these time differences are too small to be translated into perceptible changes (cf. Callender 2008).

The physical structure that plays an explanatory role with respect to our intuition of a spatially extended Now thus is the backward light cone. If we represent our own lives in Minkowski space-time by means of worldlines (this is obviously an idealization; but using world tubes of finite width does not change the general picture), this backward light cone is well-defined at each point of our worldlines—at each instant of our lives. Physics thus certainly possesses the conceptual means for explaining that during our lives we always, i.e. at each moment of our lives, have the impression to be in direct simultaneous contact with distant regions. The explanation here is a typical B-theory explanation. No special instant is singled out: the same explanatory story applies to each and every point on our worldlines.

The just-sketched explanation should give us pause. We were dealing with a very direct and strong intuition, namely that we are all living in one and the same spatially extended Now. At first sight it seems eminently plausible to view this intuition as support for the idea that the world (in the four-dimensional sense) actually is a stack of such Nows. But actually we *know* that all signals take time, and that we cannot have immediate and instantaneous contacts via a Now—so it must be that our immediate intuition is deceiving us here. Moreover, it is not difficult to find an explanation for this intuition of ours within a well-confirmed physical framework, as we have just seen. The support for the picture of the world that was based on a direct and literal interpretation of our intuition starts to evaporate once we ponder a number of uncontroversial scientific facts. In this case it is clear that although our intuition of an extended Now does reflect certain objective facts of our world, in particular the enormously high value of the speed of light, it cannot be considered to give us a literally true account of the physical world.

It follows that the strength of our intuitions cannot pretend to be an unfailing guide if we want to construct a satisfactory scientific worldview. Could not our intuition that time flows and passes belong to the same category of intuitions that have to be replaced by better informed ideas?

It is a well-known result of physiology and psychology that our awareness extends over a brief interval of time, of the order of magnitude of one second (or a bit less, depending on the person–this is the so-called *specious present*). So what we *experience* as one moment is in fact extended in physical time. The biological background of this may relate to the amount of information that is needed for action in daily life: the information contained in such a relatively short time interval is often sufficient for decisions in ordinary circumstances. If our time awareness were restricted to a single point instant, however, it would be impossible to be directly (i.e., without invoking memory) aware of change because there could be no comparison, in direct experience, of states of objects at different physical instants.

Incorporating specious presents within the B-view of time, as a description of how sentient beings become aware of time, we arrive at a picture in which there are short temporal spans of awareness strung all along the worldline of an observer. There is no preferred Now in this picture: the specious presents are all equally there in one and the same four-dimensional diagram, each one centering on its own central instant along the world line. Given their lengths of something like one second, we have to assume that these specious presents overlap (cf. Dainton 2010, ch.7).

Now, there is strong empirical evidence that differences in sensory input at different instants of time can result in an awareness of continuous motion. For example, the repetition of a brief sequence of pictures of a road, punctuated by a blank picture, creates the perception of continuous motion along the road (Mather 2010). In this case there is no "flow quality" in the input itself: there just is a brief period of time during which one picture is visible, later a similar period with a second picture, and so on; this can all be fully described in the B-manner. But our response to this sequence is characterized by a feeling of dynamism, of continuous change and flow, at each instant.

Another class of examples in which awareness of motion is generated in us concerns cases in which one and the same motionless picture is presented to us during a longer period of time. If there is *spatial* variation in the picture, a strong feeling of motion may result (a so-called "motion illusion"). Apparently, the succession of focuses of attention on different spatial parts of the picture during one specious present can in these cases produce an instantaneous and very real feeling of continuous change.

It therefore appears that our brain responds with a feeling of change, motion or flow if its specious presents are filled with sensory input that is not uniform during the physical temporal extension of the specious present. If this is right, the *flow* that we perceive and feel is a *secondary quality*: it is in us rather than in the things that are observed. According to this proposal the flow of time is in the same category as secondary qualities like color, for example "yellow": "yellow" is not literally present in observed objects, although our perception of yellow certainly corresponds to specific physical characteristics of the object and its interaction with light. It is not right to say that our perception of yellow is an *illusion*: the corresponding state of our visual apparatus may well truly indicate objective properties of the external world. Analogously, it would be deceptive to maintain that the just-suggested analysis of our perception of temporal flow boils down to the idea that flow is an illusion. Our feeling of flow corresponds to the presence of B-type variation of qualities over time; it correctly and veridically indicates the presence of this variation, that is: of *objective change*. What is illusory is only the idea that this B-type of change is unable in principle to produce the perception of flow; that for this perception to be intelligible we need something more, namely a real flow in physical reality over and above the variation in time that is accommodated by the B-series. That line of thought is similar to the argument that in order to make our perception of yellow understandable we need to suppose that yellowness is literally there in the yellow object itself.

It is important to stress that the just-mentioned strategy for explaining our awareness of time flow does not smuggle in an objective flow of time in its premises. The proposal is not that the Now slides along the instants of a specious present, passes different situations successively and thus creates the feeling associated with motion. Such an explanation would be parasitic on the assumption of the existence of real time flow and would therefore be useless. The idea rather is that different sensory inputs are co-present during a specious present, because our perception of time extends over the whole range of the specious present. It is this *joint presence* of different inputs that is responsible for our being aware of a flow.

A characteristic feature of this B-like approach is that what is explained here, our feeling of flow, does not refer to any specific privileged instant in time. All B-type explanations apply tenselessly, in this case to each and every "now" (specious present) along the worldline of a sentient being. But combined with the fact that these specious presents are perceived as undivided wholes, this account can in principle reproduce exactly what we know from direct experience: namely that we always find ourselves in a now characterized by a perceived quality of transience.

The addition to this picture of an objective, mind independent preferred Now that is really shifting in reality would not increase our possibilities of explanation. Because such a Now and its motion cannot be incorporated into the conceptual framework of scientific theory and do not play a role in the physiology of perception, these concepts cannot help us in constructing a scientifically acceptable explanation of our perception of flow, as we shall now discuss in some more detail.

1.6 Does the A-Theory of Time Add to the B-Explanation of Passage?

As mentioned in the Introduction, the main motivation for the A-theory comes from the conviction that the A-theory is able to provide a quick and convincing explanation of our experience of passage. As Le Poidevin (2009) writes:

> Now one very serious challenge to the tenseless theorist is to explain why, if time does not pass in reality, it appears to do so. What, in tenseless terms, is the basis for our experience as-of the passage of time?

to which he adds:

> Even if the tenseless theorist can discharge his obligation, the doubt remains that the tensed theorist can produce a simpler explanation of our experience.

We have already argued that there is no convincing reason to think that the tenseless B-theorist would not be able to discharge this obligation. But what about the second challenge, that even if the B-theory is successful here, the A-theory still does better?

As we emphasized, the central concept occurring in the A-theory (the flowing Now) does not make contact with the physical description of the world and cannot function in physical explanations. Even if we forget about Flow and just associate the Now with some notion of extended simultaneity, we face the serious problem that distant simultaneity is causally insignificant and therefore meaningless for the physical explanation of our experience.

B-type explanations of flow apply equally to each point along the worldline of an organism that experiences time, without singling out any preferred Now. This scientific account of experienced flow tells us that we always experience passage when there is change in the B-sense; our experience is veridical to the extent that it reflects this change. What further explanations could we wish? This question epitomizes the dilemma for the A-theory: there does not seem to be any empirical evidence that cannot in principle be accounted for by the B-theory, so that any purported additional explanatory power of the A-theory has to rely on non-empirical and extra-scientific considerations. Thus it becomes a truism that application of the A-theory cannot add to or simplify a *scientific* explanation, in contradistinction to Le Poidevin's above suggestion. A-theory explanations merely introduce additional "wheels" compared to scientific explanations, without adding new content.

Since there certainly will not be any new empirical prediction coming from the A-theory, or a new light on physical mechanisms, isn't the A-theory's function just to add words that sound reassuring to pre-scientific intuition but do not add anything to explanation? This general fear is strengthened when we look at concrete examples of A-accounts proposed in the recent literature (a sample from recent defenses of the A-theory is: Hinchliff 1996, 2000; Markosian 2004; Skow 2009, 2012; Tallant 2010; Zimmerman 2005, 2008; and further references cited in these articles).

First the A-theory faces the notorious problem of making sense of the motion of the Now at all: ordinary motion consists in spatial variation as a function of the independent variable "time", but this definition is unavailable in the case of motion of the now itself (Maudlin 2007, claims that this problem is overestimated and that time can very well flow with respect to itself; this claim is discussed and refuted in Price 2011). In the literature one finds basically two sorts of attempts at solving this problem. In the first a "supertime" T is introduced: the Now is located at time t_1 at supertime T_1. This supertime T can now serve as an independent variable: the Now is flowing at the speed $\mathrm{d}t/\mathrm{d}T$. The second, more common, attempt consists in the introduction of primitive tense operators. As pointed out by Skow (2009) however, these tense operators mimic the results of the supertime approach and are equivalent to it, even though they do not explicitly introduce supertime. For example, the following statement using the primitive tense operator "it will be the case", "*It will be the case* that the Now is located at t", is equivalent to "Relative to a point of supertime Later than the Current one, the Now is located at t" (in which Later and Current, with initial capitals, refer to instants of supertime rather than time). So these two seemingly different approaches boil down to the same thing. Both introduce elements foreign to scientific theory, either in the guise of supertime or in that of primitive tenses. To reiterate our earlier question: How might such concepts *explain* anything about time as it is known from experience? As it turns out, they are not really intended to do so. Listen to Skow (2012), who explains the achievements of the A-theory:

> B-theorists say that change is variation in ***time***. But I say:
> ***Real*** *change* is variation in ***supertime***.

So a new layer of discourse is added that depends crucially on the new primitive terms (supertime or primitive tense)—whose meaning cannot be reduced to that of concepts we already know (this is what it means that they are primitive). Small wonder that the explanations for the passage of time that are subsequently offered are completely detached from any possible scientific explanation—they are set in a world of their own. Thus, after having posited the basic explanatory principle "Change is the engine that pushes the Now into the future", Skow (2012) explains the motion of the Now as follows:

> There is irresistible pressure for the universe to change; but the universe cannot change if the Now remains at one time;
>
> The pressure forcing the universe to change then pushes the Now into the future.

This explanation is metaphysical in the uncontroversially bad sense that it does not relate to principles supported by empirical data and are fully detached from science. An account of this kind should not be accepted as an explanation of our experience of time.

1.7 Conclusion: The Passage of Time and Explanation

Basic science has no problem dealing with change: change is variation in the values of quantities during time. However, change of time itself is a meaningless concept in science: there is nothing with respect to which time could change. At first sight this appears to indicate a divide between the world as described by science and the world as we know it from experience. But on closer inspection it turns out that established scientific patterns of explanation can be applied to account for our direct experience that time passes, and that there is no reason of principle why this could not be successful. The key is to consider the passage of time as a secondary quality, in the same category as color and many other concepts that apply to our experience.

This explanation of passage as experienced does not turn our experience into an *illusion*. We experience temporal change if there indeed *is* change in time, even though this objective change does not involve a motion of the Now. In this sense our feeling of flow is veridical, in the same way as the perception of a color can be faithful to an actual state of affairs.

By contrast, accounts of our experience of passage that rely on a postulated objective flow of time have not shown that they are more than abstract metaphysical exercises without a link to what science tells us about the world. Because as they stand they do not connect to anything we know about the workings of our senses and perception, they are unable to provide explanations of our experience that can be taken seriously. In addition to this explanatory impotence, these accounts struggle with apparently insurmountable internal tensions and contradictions, as reviewed in the beginning of this essay.

References

Balashov, Y., and M. Janssen. 2003. Presentism and relativity. *British Journal for the Philosophy of Science* 54: 327–346.

Butterfield, J. 1984. Seeing the present. *Mind* 93: 161–176.

Callender, C. 2008. The common now. *Philosophical Issues* 18: 339–361.

Dainton, B. 2010. *Time and space*, 2nd ed. Durham: Acumen Publishing Limited.

Dieks, D. 2006. Becoming, relativity and locality. In *The ontology of spacetime*, ed. D. Dieks, 157–167. Amsterdam: Elsevier.

Dolev, Y. 2007. *Time and realism*. Cambridge, MA: The MIT Press.

Hinchliff, M. 1996. The puzzle of change. *Nous:* 30, Supplement; *Philosophical Perspectives 10, Metaphysics*, 119–136.

Hinchliff, M. 2000. A defense of presentism in a relativistic setting. *Philosophy of Science* 67: S575–S586.

Le Poidevin, R. 2009. The experience and perception of time. In *The stanford encyclopedia of philosophy*, ed. E.N. Zalta, http://plato.stanford.edu/archives/fall2011/entries/time-experience/

Markosian, N. 2004. A defense of presentism. In *Oxford studies in metaphysics*, vol. 1, 47–82. Oxford: Oxford University Press.

Mather, G. 2010. Motion perception. Available at http://www.lifesci.sussex.ac.uk/home/George_Mather/Motion/index.html

Maudlin, T. 2007. On the passing of time. In *The metaphysics within physics*, 104–142. Oxford: Oxford University Press.

McTaggart, J.M.E. 1908. The unreality of time. *Mind* 17: 457–474.

Oaklander, N. 2004. *The ontology of time*. Amherst: Prometheus Books.

Price, H. 2011. On the flow of time. In *The oxford handbook of philosophy of time*, ed. C. Callender, 279–311. Oxford: Oxford University Press.

Skow, B. 2009. Relativity and the moving spotlight. *The Journal of Philosophy* 106: 666–678.

Skow, B. 2012. Why does time pass? *Nous* 46: 223–242. Also available at http://web.mit.edu/bskow/www/research/whydoestimepass.pdf

Tallant, J. 2010. A sketch of a presentist theory of passage. *Erkenntnis* 73: 133–140.

Zimmerman, D. 2005. The A-theory of time, the B-theory of time, and 'taking tense Seriously'. *Dialectica* 59: 401–457.

Zimmerman, D. 2008. The privileged present: Defending an "A-theory" of time. In *Contemporary debates in metaphysics*, ed. T. Sider, J. Hawthorne, and D.W. Zimmerman, 211–225. Oxford: Blackwell.

Chapter 2
Relativity, Global Tense and Phenomenology

Yuval Dolev

Abstract An overview of the efforts of the last century to interpret relativity theory reveals that, for the most part, they concentrated on the formal and geometrical features of the theory while ignoring almost entirely its experiential side. One consequence of neglecting to examine the nature of experience is the widespread acceptance of the static block-universe picture. While many supporters of this view admit to the existence of a gap between this interpretation of relativity theory and experience, according to a suggestion made by Dieks in this volume, a closer examination shows the block-universe picture to be in perfect harmony with experience. In this paper I claim that: (a) any interpretative enterprise regarding relativity that does not include a phenomenological study is inadequate; hence Dieks' attempt to harmonize the theory with experience is commendable; (b) nevertheless, Dieks' proposal is untenable; (c) global tense and passage are irremovable from our experience-based conception of reality and must therefore figure in any interpretation of relativity theory; (d) a proper phenomenological analysis of tense and passage (based on an abandonment of the A and the B theories of time) facilitates squaring relativity with experience.

Keywords Special relativity • Simultaneity • Phenomenology • Global present • Present • Tense • Passage • Time • Static time • Block universe • Einstein • Alexandroff present • Stein present • Becoming • B-theory • Tenseless theory • Minkowski • Spacetime

---

I wish to thank the Israel Science Foundation for supporting the project of which this paper is a part (grant number 491/09).

Y. Dolev (✉)
Department of Philosophy, Bar-Ilan University, Ramat Gan 52900, Israel
e-mail: dolevyuval@gmail.com

Y. Dolev, M. Roubach (eds.), *Cosmological and Psychological Time*, Boston Studies in the Philosophy and History of Science 285,
DOI 10.1007/978-3-319-22590-6_2

2.1 Introduction

Much ink has been spilt on discussions of relativity and time, and, specifically, on the claim that relativity does away with tense and passage. Recently a new type of attack on the Global Present has emerged. The new proposal, which is due to Dieks (2006), does not consist of a new model or a new formal solution, but of a new phenomenological analysis. Placing phenomenology at the fore of the effort to understand what relativity says about time is overdue and important. While I don't share Dieks' phenomenological analysis, which I think is ultimately untenable, I am convinced that giving phenomenology a leading role in the interpretative enterprise is crucial. In the last two sections of this paper I will rely on a phenomenological study of my own to show how to be a realist about global tense and passage in a relativistic setting.

Dieks is a proponent of the so called block universe picture. But, unlike his predecessors, who recognize a tension between the tenseless block universe and experience, Dieks suggests that the block universe picture actually reflects our experience of time. His proposal offers to naturalize and render experientially normal a picture, which has often been treated with concern and suspicion precisely because it seems to be at odds with what, naively, seem to be constitutive features of reality, namely, tense and passage. Dieks strives towards a harmonization of the block universe and experience because, so it seems, he finds the discrepancy between them troubling. I think his uneasiness is in place.

Leibniz regarded sensory input as incomplete. In the centuries following him the notion that experience is reliable only in a limited manner, with true reality being hidden behind a veil that can be peeked through only with the aid of science, has become entrenched. One outcome of the discrediting of experience is the ease with which it is accepted that tense and passage are nothing but distortions that belong to *how* we experience but not to *what* we experience. But the notion that experience is essentially incomplete, imperfect, and possibly systematically misleading is deeply problematic from a scientific viewpoint. Construing tense and passage as "mistakes" of experience removes all of experience from reality to the point of leaving empirical science, science based on experience, without trustworthy input to feed on. This is all the more true with regard to the block universe picture. Standardly, this picture is understood as expressing a truth that experience hides, namely, that tense and passage are not part of the world. But discrediting experience to such an extent fatally cripples the very mechanisms through which the theory of relativity was arrived at to begin with.

I think it is the concern that, as it is usually understood, the block universe picture is self defeating in the manner just described that drives the desire to reinterpret it and reconcile it with experience. But ascertaining the existence of agreement between any interpretation of relativity and experience requires a study, not only of the interpretation in question but also of the nature of experience. Hence the essentiality of phenomenology. Dieks' endeavor to square relativity with experience should be hailed because it prompts the kind of study of experience which has, so far, been all but absent from the project of interpreting relativity, and without which this endeavor is doomed.

It is also important, however, because it occasions another critical examination of the block universe picture. The steady growth of the block universe's circle of admirers has been accompanied by a diminution of the attention given to the price tags attached to it. With time, the discomfort which the block universe initially evoked somehow dissipated and was forgotten, custom rendering normal what at first seemed strange. Dieks, however, returns to that original discomfort and aims at showing explicitly that it was without reason. He purports to establish that the experience of tense and passage does not amount to more than what one actually experiences in a block universe, and thereby exonerate the picture from the charge that it conflicts with experience.

I will argue, however, that as commendable as the ambition is, it is impossible to experientially vindicate the block universe. To subscribe to the block universe is to subscribe to a picture that hopelessly clashes with experience. Despite his efforts to show the contrary, the block universe picture cannot accommodate tense and passage as they figure in experience. If we are seeking an interpretation of relativity that squares with experience, we have to look elsewhere.

The block universe will be the topic of Sect. 2.3. In the next section, we will briefly return to the famous attempt to reconcile relativity with experience by limiting the spatial scope of tense. The upshot of this section will be that either the world is tenseless or else tense is *global*. After the rejection, in Sect. 2.3, of Dieks' stab at absorbing tense into tenseless reality, I will lay out, in Sects. 2.4 and 2.5, a way for naturally construing a global present within the framework of special relativity. Let me state immediately, however, that the alternative I will be putting forth is NOT a version of either Presentism or of the A-theory, doctrines which I hold to be as indefensible as their tenseless rivals.

2.2 First Proposal: The Spatially Restricted Present

The first rebuttal of the claim that relativity theory does away with tense and passage was Stein's, who shows (1968) that a notion of *becoming* can be integrated into Mikowski's space-time manifold. Rather than a static block-universe which does not divide into a past, a present and future, Stein offers a world in which every event has a past and a future – the event's past and future light cones, respectively, and also a present, the moment in which it becomes. In order to escape a conflict with relativity theory, the present is shrunk to a geometrical point. This way, the frame dependence of the simultaneity relationship is neutralized. If the present contracts to a point, the question which distant events are co-present with a given present event does not arise.

Almost 40 years after Stein put forth his suggestion, Pooley and Gibson (2006) develop what they call the "*Stein present*" (Savitt, who also toys with the idea, prefers the name "*Alexandroff present*"). Here's how this present is construed. First, it is conceived to be temporally extended. It is not pointlike, but rather has a fixed duration. For example, it could be 1 s long. Now, take some timelike line in Minkowski space, and consider an interval of the standard duration on that line. Following Savitt (2010) let's mark this interval as [e_o, e_1]. The interior of the intersection of the future light-cone of e_0 and the past light cone of e_1 is the *Alexandroff present* of the interval [e_o, e_1].

This conception of the present has its attractions. First, as Savitt points out, since it is defined in terms of the light-cone structure, it is not frame-dependent. More importantly, unlike Stein's present, it is not pointlike. The *Alexandroff present* is of impressive spatial dimensions, extending roughly one light second, that is, 300,000 km, around the location of e_0 and e_1. A present of such scope fits well with experience and everyday practice. We can ask what's happening now in Tokyo, for example, or on the moon. We cannot meaningfully ask what's happening now on Andromeda, but this is not a question that ever actually comes up anyway, and so nothing is lost by it being relegated to the domain of nonsensical utterances. Another, related strength of the improved model is that, according to it, the vast majority of the events that are actually important to us are spared that queer fate they were condemned to in Stein's original proposal, namely, that of becoming past after having been future without ever having been present.

However, the spatially limited present's weaknesses outweigh its strengths. Can we make sense of an invisible barrier up to which events are endowed with tensed properties and beyond which they are not? What is supposed to happen to things as they traverse from one side of this transparent film to the other? Can we even imagine that when speaking to astronauts on the moon it makes perfectly good sense to ask them "what are you doing now?", while this very question is meaningless if directed to astronauts on Mars? Obviously, much more has to be said about what presentness consists in before the idea can start making sense. A-theorists hold that when something becomes present it attains a kind of ontological superiority over things that are not present. But it seems utterly implausible that such a distinction should apply to the moon, but not to some meteorite hovering not far from its surface, or that it should "materialize" for a space vehicle approaching the moon and dissolve when it moves away from it.

And, to get petty, were there to be such a barrier, where exactly would it run? At 300,000 km? Why? Would that mean that at 300,001 km there is no "Nowness" anymore? Or does "Nowness" fade out gradually? That would give the claim that "now" is a vague predicate an original sense indeed. It may be suggested that one light second corresponds to our natural needs or to the rate of multiple brain process and, as already noted, ensures that all events relevant to our lives are included within the bubble of presentness surrounding us. Savitt points out, correctly, that the fact that an extended present's duration is fixed in relation to contingent human facts, e.g., the duration of brain processes, does not render it subjective. Evolution has created a fit between us and our environment, and that there's a match between what neurologically corresponds to "the present" and the actual present should not surprise us (just as the center of the visible spectrum coincides (approximately) with the wavelength at which the intensity of the sun's radiation peaks). Still, in light of the invisibility of the barrier of presentness, it is hard to escape the impression that this explanation is ad hoc. It's not as though we could experientially detect the barrier. We cannot approach it because it moves with us, as it were. So we conveniently speculate that it is one light second in extent because this would make the bubble of presentness large enough to render the events that matter to us present, and small enough to undercut the problems relativity raises.

Additionally, it should be noted that an extended present does not fair better than Stein's original point-like present when it comes to that bizarre consequence of Stein's view, namely, that, according to it, some events are past without ever having been present. True, the extended present releases most events we care about from this predicament. But this is no consolation at all, for we cannot make sense of an event's history consisting of a past and a future with no present in between regardless of the event's spatial location. Imagine a theory a consequence of which is that there are people who turn 50 without ever having been 40. Would we accept such a theory just because the people in question live on a very distant planet? To make things worse, on the proposal we are considering, *we* are such people for distant creatures who take us to exist outside their *Alexandroff* present.

The upshot of this is that an extended present that is not global suffers from defects that render it unsupportable. In our analysis of presentness we must choose – either every event, regardless of location, is, at some point in time, present, or else no event is ever present. That is, either there is no such thing as presentness, or else presentness is global. If it makes sense to ask what's happening now in the adjacent room, or on the palm of one's hand, or at a certain synapse in one's brain, then it makes sense to ask what's happening now in the adjacent galaxy. If we cannot meaningfully speak of present distant events, we cannot meaningfully speak of present proximate events either.

2.3 Second Proposal: Dieks' Position

Dieks' view is refreshing in that there is nothing apologetic about it. Usually, the claim that tense and passage are mere illusions is accompanied by accounts of why and how our experiences, thoughts and utterances are tensed despite the supposed fact that the reality we experience and think and speak about is tenseless. Dieks puts forth a much bolder claim: we do not need tense and passage to account for our experience. Or, more accurately, tense and passage are nothing over and above what we already get from a block universe, nothing over and above tenseless *becoming*, a notion we will explore presently.

The route to this position begins with an analysis of the phenomenological role of relativity's key term – simultaneity:

> If we decided to scrap the term 'simultaneity' from our theoretical vocabulary, no problem would arise for doing justice to our observations. This ties in with the fact that relativistic theories can be given completely local formulations—simultaneity plays no role in the dynamical laws of relativity theory. (2006, 160)
>
> Local observations—the experiences of localized observers—are invariant under different choices of the value of ε in the same way as they are invariant under different choices of coordinate systems. In particular, *it follows that those human experiences that suggest that time flows are invariant under different choices of ε*. (2006, 159)

Let me outline a train of thought which may lead to these claims. To directly (rather than via inferences and calculations) experience two events that are distant from each other as simultaneous it is necessary to experience both at once. Since they are distant from each other, one of them at least is not occurring where the observer is situated. But experience happens where the observer is, and so does not directly involve distant events. So simultaneity cannot be experienced. Take for example someone who sees two flares that are distant from each other ignite together. The experience takes place not where the flares are igniting, but where the observer is located. Whether these events are simultaneous or not is something that is not given in the experience. The observation is local, that is, distant from the location of the events experienced, and how it is situated with respect to them temporally is something that needs to be calculated. Strictly speaking, then, simultaneity does not figure directly in experience and hence "the term 'simultaneity' can be scraped from our theoretical vocabulary, without any injustice being done to our observations". An immediate consequence is that any way we set ε and any choice of coordinate system we make will be compatible with our observations and experience.

Note, however, that the above train of reasoning holds true in Newtonian physics as well. There too it is possible that two events will seem to be simultaneous even though they are not, e.g., if one of them occurs much farther away from the observer and earlier than the other (and vice versa, events that are simultaneous may be experienced as though they are not). Relativity introduces the dramatic addition that the result of the calculation determining whether the two events are simultaneous will also depend on the choice of a frame of reference. But the above claim about observing simultaneity does not apply solely to relativity theory. In fact, the claim that simultaneity does not play an experiential role is, in essence, not a scientific but a phenomenological one. It is possible to speculate outside modern science whether two distant events that appear to be simultaneous really are, and to base this speculation on the hypothesis that it takes time for information from the events to reach our sense organs.[1] Whether or not simultaneity figures in experience is not a question to be decided by physics, but by reflection on experience. And note that in arguing that it does not, Dieks is not relying on physics, but is making a phenomenological claim as part of his effort to interpret a theory of physics.

This phenomenological claim bears immediate consequences for tense. For the presentness of a distant event is bound up, so it seems, with its simultaneity with

[1] Already in the fifth century BC Empedocles argued that light travels at a finite speed.

what is happening here now. Graphically, distant present events lie on a hyper-plane of simultaneity together with local present events. Scrape away the simultaneity relation and an essential feature of distant (and *ipso facto*, of global) presentness is undercut. Of course, it is this entailment that Dieks is after.

But the claim that simultaneity does not enter experience is not as innocent as the above reconstruction may make it appear. That it makes no observational, experiential difference which event is happening now on Alpha Centauri may be true. But is it also the case with respect to events in Tokyo, or in the adjacent room? Could simultaneity and tense be scraped from time as it is actually experienced locally? In light of the conclusion of the previous section the answer had better be "yes", for recall that a forced choice was derived there: tense is either global or else non-existent, even locally. And indeed Dieks is unequivocal. The abolition of simultaneity and tense is total, and pertains to the most local events there are – our own experiences:

> It is the purpose of the four-dimensional spacetime picture, which the block universe is, to represent *all* events that actually take place in the universe, complete with *all their properties and mutual relations*. An adequate block universe representation therefore also contains all events in the lives of individual human beings, with all the impressions and experiences that (partly) constitute these events. For example, that I now remember past events and do not yet know much about what is to come is part of my experience at this instant of my life and should be part of the four-dimensional picture; the same applies to my conviction that exactly *now* it is now. All actual events, experiences and intuitions must be there in the block representation, exactly at the spacetime position where they actually occur. **So there cannot be any conflict between experience and the block universe**. (2006, 169. Last emphasis is mine)

Obviously, then, according to Dieks simultaneity does not need to be part of our conception of reality, not only on cosmological scales, but also on the scale of everyday experience, which, after all, is nothing but a subset of the events that make up the block universe. And, to repeat, the reasoning is this: experience is local. It does not matter whether we are looking at a distant star or at a fly standing on the tip of our nose: the visual experience is spatially separated from the event experienced, and the simultaneity (or lack thereof) of the two does not enter into the experience.

The elimination of simultaneity, not only from our theoretical vocabulary but also from our phenomenological account of experience, is Dieks' negative thesis. Dieks then moves on to offer an alternative. Simultaneity and tense are replaced in Dieks' scheme of things by the notion of *becoming*, which according to him does all the work required, insofar as accounting for experience and theory are concerned.

> Events come into being by occurring, by happening; what else could their coming into being be? … In the block picture it is recorded for each actual event *that*, and where/when it occurs… 'coming into being' means the same thing as 'happening'. … There is no need to augment the block universe in any way … becoming is nothing but the happening of events, in their temporal order. (2006, 170–1)[2]

---

[2] Let me mention in passing, that even though the replacement of tense with this notion of becoming should supposedly be music to the ears of new B-theorists, a further examination reveals

Dieks' notion of *becoming* is simple, parsimonious, and it eliminates with one stroke the conceptual conundrums given rise to by relativity. But it can sustain a viable interpretation of the theory only if it can deliver on its promise and replace simultaneity, tense and passage in our description and analysis of experience. Well, can it?

I will turn to this question momentarily, but before doing so I wish to assess Dieks negative thesis concerning the supposed redundancy of simultaneity. Can we seriously consider the denial that during a hockey match the forward and the defenseman are pocking at the puck at the same time? Or that during a concert the arms of the orchestra's violinists are moving together? Or that in an accelerator the magnetic field is oscillating at the same time that the particles are making their rounds through the tunnel? If there is any issue with the simultaneity of these pairs of events it is, to the contrary, due to the foundational role simultaneity plays in such cases. Far from not figuring in experience, simultaneity belongs to the perhaps ineffable substructure on which experience is conditioned. If we cannot say in such cases that the events in question are simultaneous the reason is that it is so obvious it is not clear what is being asserted. The difficulty with such assertions is akin to that encountered when denying that this is a hand, or that the earth has existed for more than 5 min. The issue is not so much one of falseness as of meaninglessness. The demons of skepticism can pester us with various scenarios. But indulging in such fantasies shakes the very foundations on which the possibility of doing science relies. It is probably not possible to fantasize a world devoid of simultaneity relations, but were it possible, the world fantasized would be one in which science, as we know it, would not exist either.

Now, the last thing Dieks intended was to put forth a skeptical argument, or one that raises foundational issues concerning science. To the contrary, his aim is to reconcile relativity with experience, not to raise questions about the reliability of experience. But in that case it must be recognized (a) that the simultaneity of events is part of the input science feeds on, and (b) that what we learn from relativity is that simultaneity is not invariant under change of reference frames and not that, e.g., it can coherently be denied that the violas and the violins are playing together, that is, simultaneously.

Science tells us that simultaneity is more intricate than was thought. Early science revealed that light travels at a finite velocity, from which it followed that very distant things become visible after, sometimes long after, they occur. That already complicates the handling of simultaneity. And special relativity brought with it the discovery that simultaneity is a frame-dependent relationship. Neither of these two complications, however, undercut the basic role simultaneity has in our understanding of things. The notion of simultaneity is an irremovable component of the

implicit themes that they will find disturbing. After all, simultaneity figures crucially in the semantic apparatus the new B theory relies on. The tenseless truthmakers of tensed truths can fulfill their task in virtue of their being simultaneous with tokens of these tensed truths. Taking away simultaneity is therefore crippling for B-theorists who do not want to deny that tense figures in experience and language.

conditions for science, even if this very science uncovers startling complexities with respect to it.

Now on to Dieks' positive proposal, viz., that a notion of "becoming" can replace the distinction between the past, present and future. Recall that Dieks' aim is not to establish that the block universe is logically or scientifically or conceptually sound, but that it is supported by experience, or at any rate, does not clash with it. This is an unorthodox position. There's widespread agreement that tense and passage of the kind absent from the block universe (in a minute we will say what these are) are not eliminable from how we experience, think and speak. The staunchest supporters of the B theory acknowledge the gap between experience and the view of reality afforded by their theory. Indeed, the so called "new" B theory is different from its predecessor precisely in that rather than trying to explain away the tensed features of language and experience, it retains them by providing for them tenseless truth makers.

What's truly exciting about Dieks' approach is that it treats all this labor not as superfluous but as misguided. According to him there is no gap between reality and how it seems to us to be, and so no conceptual analysis is required for the sake of bridging any gaps. The aim of conceptual analysis is rather to flesh out and study what experience actually tells us about reality. And the key is the notion of "becoming". Becoming is supposed to pack within it everything necessary for describing tense and passage as they figure in experience.

Unfortunately, this notion does not stand up to the task. The kind of study of experience Dieks' thesis prompts mercilessly points to the gap Dieks attempts to deny. To begin with, the very fact that it requires an argument to show the existence of harmony between the block universe and experience already establishes the absence of such harmony. We do not need to sweat with respect to the fact that the sun rises to the east, that it takes time to get from home to work, and that the bus's wheels are round – these, and countless such facts are in complete match with how we experience them. It would take a bit of clever argumentation (again, of the skeptical variety) to get us to suspect that there's a gap in such instances between reality and our experience of it. In contrast, the claim that the universe is a block in which events' becoming is all that passage and tense come to, rather than simply describe what we always thought was the case, startles us.

It startles us because reflection on experience reveals tense and passage are more than what can be unpacked from "becoming". Arriving at this "more" in itself requires some unpacking. We start by noting, trivially, perhaps, but not insignificantly, that the events that make up reality are experienced, thought of and spoken of as *themselves* tensed, as being past, or present or future. It is not that tense is something that predicates them *as they are experienced*, rather, it is something that predicates them as they are in themselves, or at least that is how we experience them to be, as tensed irrespective of experience. I am not claiming that this proves tense is a property of events, I am stressing however that we experience tense as a property of events and not merely of how events are experienced.

Consider Max, who at *t* thinks to himself "It's been a long time since my last visit to Paris". In Max's ruminations, his Paris visit *itself* is past, it is that event, the visit, that carries the property of pastness. Pastness qualifies the event, not Max's remembering the event. But on Dieks' phenomenological account of "becoming", "Events come into being by occurring, by happening; what else could their coming into being be?". Max's remembering is occurring, happening, and in it the visit to Paris is "painted" as past, but its pastness is no more than a sentiment accompanying the remembering. Pastness is not a feature of the event, rather it is a component, one among many, that make a certain mental event, Max's remembering a visit to Paris, into what it is. In the block universe the only temporal relations between events, e.g., between Max's visit to Paris and his reminiscing about that visit are relations of succession. An event being past is not part of the temporal relations that obtain.

Indeed, according to B-theorists Max's visit to Paris is not past, since nothing is past (or present or future). But Max apprehends it as past. So Max's remembering the visit involves a kind of *misrepresentation*. The event is not past, yet it is apprehended as past. Most B-theorists see no problem with this state of affairs. A mistake occurs only when a tensed metaphysics is read off from tensed beliefs, which in themselves, B-theorists haste to stress, can be true and useful. The mistake is averted, however, once, with the aid of the appropriate semantic theory, tesneless facts are shown to constitute the ground for tense beliefs and truths.

There is, then, a crucial difference between Dieks and other B-theorists. Both agree that tensed sentences and beliefs do not express and do not attest to the existence of tensed facts, at least those that go beyond what can be unpacked from "becoming". But B theorists concede that experience and language mislead us, initially, at least, into believing in the existence of tensed facts, while Dieks states that a close look at experience shows it does no such thing. Experience, according to him, does not involve any kind of misrepresentation. In other words, the experiential gap between the tenseless facts of reality and the tensed manner in which we experience reality and speak of it, a gap new B theorists account for with a semantic theory which explains how tensed beliefs are made true by tenseless facts, is denied by Dieks.[3]

Here's, however, why he is wrong. Not only are tenseless beliefs about succession insufficient practically (knowing that the train to Boston leaves 15 min after the train to NY will not get us to either train on time if we do not know what time it is *now*), they cannot capture the omnipresent and hyper-significant *emotive* role of tense, e.g., the massive emotive shift we undergo during the few minutes in the course of which a future bungee jump becomes present and then past. As B-theorists are quick to admit (and explain) the shift has to do with *tense* – anxiety raises as the jump "approaches" and dissolves once it is past and over. Never-changing, tenseless, successive relations leave us unperturbed. It doesn't matter to us that a bungee jump is (always was, always will be) before lunch. It matters to us whether it is before or after this *present* moment. Being past (or future) is, phenomenologically,

[3]There are other instances of this difference between Dieks' and B-theorists. For example, some B-theorists acknowledge that there's a gap between how motion is experienced and what motion in the block universe really is (cf. Paul, L.A., "Temporal Experience", Journal of Philosophy CVII (7): 333–359 (2010)).

more than merely being before (or after) something. Indeed, it is this failure of reducing tense to succession that was the impetus for devising the new B-theory, in which tense is rehabilitated as an irreducible feature of our conception of (tenseless) reality.

But we're not through yet. It may be replied that "becoming" is not designed to rescue that hopeless program of reducing tense to tenseless relations. "Becoming" contains more than just succession. It consists of succession *plus* the uniqueness which events that are "becoming" enjoy with respect to those that precede or follow them. It may be suggested that the occurring of something impregnates it with emotive richness which it lacks when viewed from a temporal distance. "Becoming" gives the event that becomes its livelihood, as it were. The bungee jump is experienced in one way when it becomes, and differently (or, strictly speaking, not at all) when it is merely anticipated or remembered. It's like, to revert to old analogies, the livelihood that the sentence "I am hungry" carries when it is uttered by *me*, or the special status a place has for those who can refer to it as "here". In the case of time, from its own temporal standpoint, as it were, each event, one's experiences included, is special in that only it is happening, occurring, only it is "real and alive". This proposal augments succession with an additional element, a distant echo of the A-theoretic notion that the present moment is unique, and elevates "becoming" into something more than mere succession. If successful, this move renders the block universe a place worthy of the colorful drama reality as we experience it is.

But the echo is too distant. It makes the becoming event unique in the same way Max is unique in that only he can refer to himself with "I", that is, if anything, in a purely subjective way. But, phenomenologically speaking, present events are unique in a non-subjective way. There is a dis-analogy between "now" on the one hand and "I" and "here" on the other, the difference consisting in that we regularly speak of other persons and other places as co-existing in time with us and with the location we refer to as "here" but we never claim that events that are not present co-exist in time with those that are. So unless we beg the question by assuming some kind of tenseless existence which all events share regardless of their temporal locations, we cannot in any context or way say of events that are past or future that they exist. We speak of them as having existed or as going to exist, but never as existing. It's this feature of tense that generates the popular belief that "only the present exists".

Note, importantly, that the observation just made *does not* entail the metaphysical thesis that only what is present exists, or any of the tenets of the A-theory or of presentism. Indeed, I find this theory to be as untenable as the rival B-theory. What remains the case, however, is that while different people and places coexist in time, events that are not present do not coexist with those that are, a fact that phenomenologically sets present events apart from those that are not. This phenomenologically undeniable distinction of present events (which, again, has nothing to do with the ontological distinction A-theorists attribute to present events) is absent from Dieks' proposal. It's not, as Dieks would have it, that all events are "becoming", each at its temporal location. Rather, some are "becoming" but others "have become" and others yet "will become". This phenomenological uniqueness of present events is left out of the frozen block universe's view from nowhere.[4]

It will immediately be objected that the above also begs the question in that it assumes a tensed language. Answering this charge will drag us too deeply into the A/B theory debate, which is not the topic of this discussion. The point pertinent to us is that imbuing "becoming" with the only kind of "uniqueness" which is possible within the context of the block universe, namely, the temporal, tenseless equivalent of "I" or of "here", does not enrich this term enough to make it encompass tense as we know it from experience. What is necessary for a full and faithful description of experience is not a tautological tenseless sense of "becoming", one in which every event becomes when it occurs, but a tensed sense of "becoming", according to which events that are *not* present are *not* becoming (and again I must stress that this tensed notion does not have to be in the vein of the the A-theory, that is, ontological).

To sum up, adherents of the block universe need to make a case for it precisely because the essence of the temporal aspects of experience is absent from it. And "becoming" is too weak of a notion to fill in for what's lacking.

I said above that Dieks' view is refreshing in that rather than engage in acrobatics in order to reconcile the block universe with experience, it claims that experience and the block universe are already aligned. Dieks' maneuver fails. But concealed in it is a more general moral which I think is valid and should be voiced loudly, namely, that we ought to be unapologetic about experience. Rather than treat experience as confused and inferior we should regard it as, in general, accurate, trustworthy and crucial for science. Interpretations of theories that do not clash with phenomenology are not a luxury but a necessity. No scientific theory can be the basis for a claim which, if it were true, would render the generation of that theory impossible. But a theory that sweepingly undercuts the reliability of experience does just that.

On the standard understanding of the block universe relativity theory reveals a momentous gap between experience and reality. The gap is not merely between the invariance of experienced simultaneity and the frame dependence of simultaneity in relativistic situations. It's a comprehensive, ubiquitous gap, between all of reality, which is frozen and static, and all of experience, which is dynamic and tensed. Dieks seeks a different understanding of the block universe picture, one which reconciles relativity with experience, one in which, indeed, experience corroborates the block universe. I have argued that while the quest for reconciliation is appropriate, the hope that it will be achieved within the framework of the block universe must be abandoned. Later I will outline an alternative approach for harmonizing relativity with experience.

---

[4]The amended proposal – succession plus uniqueness of the becoming event – is challenged by further queries: why does the temporal distance to an event matter? Why do we get more nervous the more the moment of the bungee jump "approaches", and why does it matter whether it is before or after where we are now? The asymmetry between "before" and "after" remains a mystery, certainly for those who do not buy into collapsing temporal order onto causal order.

Let me close this section with a brief comment on Stein. Stein's classic work is the soil from which both the spatially restricted present and the notion of "becoming" grow (the notion figures centrally in Stein's own work, of course). But the spatially restricted present is underpinned by a desire to reintegrate the dynamism of passage into Minkowski space-time, while Dieks' "becoming" is invoked in an attempt to legitimize the static block universe. How is it that these two opposed views find their origins in Stein? I think the explanation is this. Stein felt that the conclusions derived by Putnam (1967) and Rietdijk (1966), in which tense and passage were removed from relativistic reality, must be rebuffed, precisely because the manner in which these conclusions undercut the trustworthiness of experience threatened to bring experience-based science down. So Stein turned to the notion of becoming and expounded it geometrically, thereby reinstituting, supposedly, tense and passage back into the picture. However, the notion he offered was in fact itself tenseless. Stein's becoming, like Dieks', is viewed from nowhere, and fits perfectly with the B-theoretical recognition that, tautologically, events occur (tenselessly) when they do. Stein too failed to appreciate that to interpret relativity what is needed is not more geometry but more phenomenology. A more careful phenomenological inquiry would have revealed that tense is not about the "ontological superiority" that generates the difficulties in the relativistic setting, and would have led to a conception of tense that can be fitted rather naturally into such a setting.

2.4 Distant Events

Here's our dilemma once again. Either tense and passage are part of the fundamental structure of reality or else our apprehension of reality is, as Leibniz said, incomplete and confused. Dieks tries to evade this dilemma by suggesting that tense and passage belong neither to the fundamental structure of reality nor to how we experience reality. More accurately, according to him the notion of "becoming" can facilitate an adequate description of both reality and experience. However, in the previous section it was shown that this claim is indefensible. I believe that, after almost a century of struggling to squeeze dynamism out of the frozen block universe or inject dynamism into it, we ought to archive the block universe and return to a natural conception of reality, one which takes tense and passage to be constituents of its basic structure.

Let us then accept as a statement about reality that every event, *anywhere* (cf. Sect. 2.2), is either past, present or future. Someone is sipping coffee now in Tokyo, Armstrong's setting foot on the moon is now 46 (or more) years past, and the next time Venus will come between the earth and the sun will be in approximately 100 years. These tensed properties belong to the events in question, not our apprehension of them, no more and no less than the spatial locations and velocities of the objects figuring in them. But let's move over to cosmological scales and to situations which manifest the bizarre states of affairs entailed by relativity, those that generate the thought that relativity effaces tense and passage from our understanding of reality.

Max is pacing back and forth in his room wondering "What is happening now on Andromeda?"[5] Due to the immense distance of Max's room from Andromeda (2.5 million light years), the events Max is thinking of as occurring now while he paces eastward are separated from those the word "now" picks when he is pacing westward by many years. If Janet is in the room as well, pacing in opposite direction to Max's, they will disagree as to what is happening now on Andromeda, or, more accurately, on where in time the events happening now on Andromeda are located. This consequence assumes that we identify their "presents" with the hyper plane orthogonal to their world-lines (a point we shall return to shortly). But even if Janet and Max are walking together in the same direction a conflict may arise if they choose different values of ε, that is, pick different foliations out of the infinitely many foliations compatible with the empirical data they share. How should this outlandish situation be treated?

To understand tense in a relativistic world let us first study what it means to attribute tensed properties to distant events in non-relativistic situations. What do we mean when we ordinarily think or speak of a distant event as present? Obviously, the notion of a distant present event is rather complicated and is attained after more basic temporal concepts are in place. Of all the components that later make up one's notion of time, the one given most immediately and directly, perhaps the only one given directly in experience, is that of the presentness of the experienced. The events we experience are invariably present (the few exceptions are distant stars).[6] Building on this basis, one's picture of time is expanded, both "vertically", to include events that are later and earlier than those present, and "horizontally", to include distant present events and relations of simultaneity. How do these expansions come about? Obviously not through direct experience, for the events figuring in the expansions are not available to experience – they are distant either in time (later or earlier than the present one can experience), or in space, or in both. For the purposes of the present paper we focus on horizontal expansion. I suggest that in this case (but also in certain aspects of vertical expansion) "counterfactual experiences", experiences one would have had were one located at the distant location in question, play a central role.

An infant, perplexed by the absence of his older sister, is told that she is away skiing. If he doesn't understand, it can be helpful to show him a video clip of someone skiing and tell him – "that's what your sister is doing now; if instead of being here we were there, that's what we'd see her doing". Current technology actually enables showing him his sister skiing in real time, and saying the same – if we were there we'd see her with our eyes, not on a screen. Thus, a counterfactual experience, the experience one would have had if one was at the distant location, is an important

[5] NASA engineers actually find themselves in such situations, e.g., with missions to Mars, in which signals from critical stages of the mission may travel up to 10 min before reaching mission control.

[6] This is something most B-theorists would agree to. Mellor, for example, claims that in the process of forming our conception of time tense has priority over tenseless relations. Of course, they interpret "presence" in a tenseless way.

component in the introduction of the notion of a distant event to one not familiar with it. But the crucial part that must be highlighted is that the counterfactual experience is one the person would have had had she been there *now*. But what does that mean? How does tense enter the counterfactual, imagined experience? What differentiates imagining (or seeing on a screen) an event at a distant location occurring *now* from imagining it having occurred earlier or about to occur later than now (an option that applies only to the imagination and not to the screen)?

I know what it is to have been elsewhere an hour ago, and also what it means to say where I'll be an hour hence. An hour ago I was at home. It took me an hour to travel to where I am now. And if I wanted to return home it would take me almost an hour to get back. If I could travel with a jet, the time it would take me to get home would drop to under a minute. That would still not enable me to be there now.[7] To be there now I would have to be transported there with infinite velocity. So let's amend the above. Speaking or thinking of distant present events involves the counterfactual experience we would have were we transported to the distant location at an infinite velocity.[8]

Some clarifications are due. For many distant events, transportation at finite velocities would still get us there on time to witness them. We can walk 30 min and still make it to the spring carnival at the school. But the shorter in duration the event happening "there" is, the faster we need to get to it before it is over. And by letting this duration shrink to zero we realize the velocity of transportation has to be infinite. A little reflection shows this is true also for events that are not of short duration. Transportation at finite speed will always get us there with a delay with respect to what is happening *now*, strictly speaking, which even if infinitesimal, is still a delay. Hence, the notion of the presentness of a distant event is tied up with transportation at infinite speed.

What about events that, for whatever reason (because they involve objects that are too small, or because they are too brief, or require vision outside the visual spectrum, etc.) cannot be experienced at all? These do not challenge the above scheme. Once we are in possession of the notion that something is happening now we can apply it to events that cannot be experienced, whether they are present or distant.

We similarly explain cases in which we are thinking of distant events without any imagined counterfactual experience taking place. Hearing on the news that a storm is raging in NH does not have to be accompanied by any visual image. Once we have available the notion of a present distant event, which is originally formed through cases involving counterfactual experiences, we can readily understanding this bit of news.

[7] The present is being treated here as though it is pointlike. Elsewhere I argue that this is a problematic assumption, but problems concerning this assumption do not effect the current argument, for the purposes of which it is useful to think of the present in this way.

[8] With its invocation of counterfactuals this is a rather complicated form of connecting to the temporal properties of distant events. But then again there is no reason to suppose that creatures lacking this kind of conceptual machinery ever engage in speculations about distant events.

Let's take stock. On this scheme just outlined tense plays a constitutive role in the process by which we develop our conception of time. The first step consists in apprehending, albeit, tacitly, the presence of the experienced. As our understanding of temporality develops it comes to include events happening at a distance. Here too the presentness of the experienced plays a constitutive role, even though the experience in question is a counterfactual one – the one had were one transported "there" at infinite speed.

It follows from this phenomenological analysis that distant events that are present are those we'd encounter had we been traveling at infinite velocity. In a graphic representation of time, co-present with one's present experiences are those events that are reached if one travels at infinite speed, namely, those occupying the plane orthogonal to one's world-line.

This indeed is how co-presentness was conceived in classical physics. In view of the fact that this notion of co-presentness is constitutive of how we grasp time, we mustn't abandon this conception, unless it becomes absolutely impossible not to do so. My claim is that relativity gives us no reason to question or dispose of this notion of co-presentness.

2.5 Tense in a Relativistic Setting

The above discussion refrained from asking what presentness comes to – what it is that differentiates present events from those that are not. This issue is at the heart of the heated debate in the metaphysics of time. But this is not the place to elaborate on any of the answers to it. The purposes of this paper are served by the exposure of the central role that tense plays in the formation of our conception of time, and specifically, of our notion of distant events. It must be emphasized again that when speaking of tense we are speaking of properties of events. We experience events as *themselves* present (or past, or future). While this fact is recognized by most B-theorists, who give it a tenseless interpretation, some B-theorists may object to parts of the above scheme. The aim of this paper, however, is restricted to showing that no such challenge comes on the grounds of relativity theory.

Let it be noted, first, that other places, even if they are very distant from us, are not the wild-west, where everything goes. Frogs are not turning into the number 2 on Andromeda, and if there is water somewhere in that galaxy then it does not become gold or begin to sing when boiled. Sure, claims about events that are space-like separated from us cannot be refuted or confirmed empirically. That doesn't mean anything can be hypothesized about such locations. If there's a sun in Andromeda with three small planets and three large ones orbiting it, it cannot be speculated whether this solar system consists of seven planets. To be sure, the problem with such a speculation would be not that it was evidently false, but that we would not know what meaning to attach to the words of someone raising it. Distance, experiential inaccessibility, in and of itself does not convert nonsense into sense.

The mere fact that events are space-like separated from us does not license us to attribute to them features that our language does not know how to express, or our thinking to think.

That is true specifically of tense. Andromeda is no different than Tokyo or the tip of one's nose. If, as most B-theorists acknowledge, language cannot be de-tensed then we cannot speak and think of things tenselessly, whether they are on the palm of our hand or on the surface of a remote planet. The boundaries beyond which the meaningfulness of our utterances begins to falter do not have spatial constraints. In short, the mere fact that Andromeda is very distant, and that it is impossible to empirically verify or refute statements about events there, does not by itself sanction the claim that events on Andromeda lack tensed properties. A corollary to this is that while it is impossible to say *what* is happening on Andromeda now (or, for that matter, on Jupiter's moon Pan or at the center of the earth and even at most locations on the earth's surface), there's nothing stopping us from knowing *that* something is happening now on Andromeda.

But what could attributing tensed properties to events on Andromeda come to, and how should the twofold ambiguity (foliation and frame dependence) relativity introduces be handled? The results of the previous section answer that. As we saw there, to temporally identify these events all we need is the hyperplane orthogonal to the point we are occupying now on our worldline. This notion of global presentness follows directly from the way temporality is given to us in experience.

What relativity reveals is that like mass, size, shape, temperature, the present is frame dependent. That is surprising, but, as in the case of the other properties just mentioned, being frame dependent does not amount to being "not real" or "arbitrary". In other words, tense cannot just be picked out at random from the many foliations that are empirically compatible with one's observations.

Returning to Max and Janet, their situation seems much less daunting than at first. Events on Andromeda, like events anywhere, are past or present or future. The ambiguity concerning which event is occurring there now that arises from the supposed existence of non-standard choices of ϵ is gone. And the one that has to do with the frame-dependence of simultaneity is absorbed – tense is frame dependent. In their musings about what's happening now on Andromeda Max and Janet speak of different events. But that's no more surprising than the differences between the results they attain when measuring the mass, spatial dimensions or velocities of hypothesized objects on Andromeda. The tenses, recall, are properties of events, properties which like other properties, are not invariant.

The view that the present is frame dependent and not arbitrary is not new. But what I propose differs from previous suggestions in two essential ways. First, the argument I put forth is not formal or geometrical, and so has nothing in common with, e.g. Malament's (1977) famous argument. Second, as already stated above, my suggestion is as distant from Presentism as it is from the block universe. It by no means relies on or entails the thesis that only what is present exists, and so has nothing in common, e.g., with so-called "surface presentism" discussed by Hinchliff (2000).

To sum up, though the choices of a foliation and of a frame of reference are experientially irrelevant, they are not arbitrary. Their experiential "transparency" does not lead to the conclusion that tense and simultaneity do not figure in how we experience the world. Rather, it forces us to engage in a phenomenological study of tense and simultaneity, a study which, among other things, shows them to belong to the very setting in which we experience, conceptualize, and do science.

One last question must be addressed before we conclude this section. If such a manner to square tense with relativity is available, why have so many endorsed the claim that relativity does away with tense and passage? There are two reasons for this, which may be related. The first is the authority of Einstein, who on various occasions defended the block universe view. Einstein, and more so his followers, crushed any attempt to defend the reality of tense. The second has to do with the A/B theory debate. In a nut shell, the heart of this debate concerns *ontology*. For A theorists present events enjoy an important ontological superiority with respect to those that are not present. For B theorists all events are on an ontological par. Relativity clashes with the A theory's ontology. It is indeed implausible that as Max is pacing back and forth in his room events on Andromeda go in and out of existence, which happens (regardless of which foliation we choose to identify the present with) if we accept that "only what is present exists". Many take the incompatibility of relativity with the A theory to entail the correctness of the B-theory, or of the block universe.

In my previous work I have argued that both the A and B theories must be abandoned, and that the ontological analysis of tense they are committed to must be superseded by a phenomenological inquiry. This inquiry, I tried to show, yields a realism about time that is not fleshed out in terms of reality claims, the kind of claims which capture the tenets of the A and B theories ("only what is present exists"; "everything exists, tenselessly"). I cannot here present the non-ontological conception of tense I offer. The point I want to make here is that once the ontology of the A theory is rejected, the argument from relativity to the block universe is undermined. The argument is a negative one – it assumes that there's a forced choice between the A and B theories, and that therefore if the A theory is not tenable then the B theory is correct. Once the rug is pulled out from under this forced choice, the argument loses its validity, and the door is opened to a conception of relativistic reality with tense and passage as part of it.

2.6 Conclusion

There's a sense in which the debate concerning relativity and tense is "academic". We cannot say of any *concrete* event on Andromeda that it is present because we don't know of any concrete event on Andromeda – no concrete event dating from the last 2.5 million years is experientially accessible. Janet's and Max's musings about current events on Andromeda are utterly "philosophical". When we insist, as I have been insisting, that it is meaningful to state that something is happening now on Andromeda, the insistence pertains to a statement that is void of any informational content. Why then are we interested in the question of global tense and passage?

This is a question all must contend with, not just those who hold that tense is global. Champions of the block universe are also obliged to answer it, for the claim that time does not flow on Andromeda is just as "academic" and just as devoid of any empirical and experiential significance as the contrary view, for which I have been arguing.

One reason distant events interest us is that eventually, given enough time, what happens at any location X may become relevant to occurrences at any other location Y.[9] A collapse of the Tokyo stock exchange sends shock waves across the world's financial markets, communication devices propagating these waves at the speed of light. If Neptune were to explode now, in a few hours the earth would have been showered with the first products of this explosion.

But the question of global tense is interesting and important for different reasons. First, whether or not time flows seems to have, let us call them, existential implications, which obviously cannot be broached here but must not be underestimated. Accepting the block universe picture globally means accepting it ubiquitously. And this is bound to have an effect on how we perceive our place in the world. Reflect for a moment on the recurring claim that our attitude towards our own death would change radically were we to acknowledge that passage is an illusion.[10] So we want to know if time flows on Andromeda because if it does then it flows here and if it doesn't then it doesn't flow here either. Those who brush such considerations aside dismissively ought to recall that they were at the center of the debate between Einstein and his opponents, among whom were found many major physicists and mathematicians that insisted on the significance of "lived time".

Second, the issue of global tense has an important place in any study of the assumptions concerning the unity of nature. The argument of this paper entails a unity principle, which bears, I believe, close kinship to one of the pillars sustaining relativity theory itself, namely, the principle of relativity. The principle of relativity states that the laws of nature have the same form in all frames of reference, and specifically at all locations and times. But a close look uncovers that the principle of relativity in turn presupposes another principle, namely, that experience has the same form regardless of frame of reference. If the nature of experience itself were to vary with shifts in frames of reference it would be impossible to give meaning to the principle of relativity, which, after all, among other things, says something about what we should expect to observe and experience under different circumstances. The uniformity of the forms of the laws of nature presupposes the uniformity of experience. True, the principle of relativity does not mention, not even implicitly, tense and passage, and this is probably part of the reason why the removal of tense

[9] Leaving aside horizon considerations.

[10] As is well known Einstein's assertion that the distinction between the past, present and future is merely an illusion appears as part of the consolation Einstein offered Ms. Besso following the death of her husband.

and passage from our picture of reality meets little resistance. But, if assuming the universality of the laws of nature entails assuming the universality of the structure of experience, and if experience is tensed, then tense is woven into the principle of relativity.

Thus, the philosophical analysis of tense and relativity uncovers unity assumptions that, far from being a philosophical extravagance, are part of the background that makes science possible. Tense may not figure in the facts of reality that can be gleaned from scientific investigation, but it certainly figures in the facts of the reality within which this investigation takes place. Affirming time's passage on Andromeda, no more and no less than any other statement about that distant galaxy, says something about the spatio-temporal structure of the universe, about time being the dimension of change, and change having to do, as McTaggart (and Leibniz, and Whitehead, and James, and the list goes on and on) adamantly insisted, with tense, with future things becoming present and then past. It tells us that even though physical reality is replete with surprises, they too happen in an arena that is in some fundamental aspects familiar.

This result downsizes somewhat the scope of the revolution relativity theory has brought to our conception of time. Time dilation is startling enough. But the more impressive abolishment of tense and passage attributed to relativity theory is not in fact entailed by it. Still, our lesson from relativity is an important one. That even after relativity tense and passage continue to constitute a fundamental feature of time, or that, at any rate, it will take more than what physics currently offers to establish the contrary, is in itself a significant discovery about how deeply tense is rooted in our conception of reality.

References

Dieks, D. 2006. Becoming, relativity and locality. In *The ontology of spacetime*, ed. Dennis Dieks. Amsterdam: Elsevier.

Gibson, I., and O. Pooley. 2006. Relativistic persistence. *Philosophical Perspectives* 20: 157–198.

Hinchliff, M. 2000. A defense of presentism in a relativistic setting. *Philosophy of Science* 67: S575–S586.

Malament, D. 1977. Causal theories of time and the conventionality of simultaniety. *Noûs* 11: 293–300.

Putnam, H. 1967. Time and physical geometry. *Journal of Philosophy* 64(8): 240–247.

Rietdijk, C.W. 1966. A rigorous proof of determinism derived from the special theory of relativity. *Philosophy of Science* 33(4): 341–344.

Savitt, S. 2010. Relativity, locality and tense. In *EPSA philosophical issues in the sciences*, 211–218. Dordrecht: Springer.

Stein, H. 1968. On Einstein-Minkowski space-time. *The Journal of Philosophy* 65(1): 5–23.

Chapter 3
Why Presentism Cannot Be Refuted by Special Relativity

Yehiel Cohen

Abstract It has been argued that Special relativity with its most striking feature, namely that the definition of the present depends on a choice of an inertial frame, actually refutes *presentism* – the metaphysical view that only present events are real. Contrariwise, it has also been argued that the notion of the present in a relativistic setting is not a matter-of-fact but established by convention and hence devoid of metaphysical interest. In this paper, I contend that, considering Einstein's philosophical motivations for introducing special relativity, both of these assertions are wrong and that the pre-relativistic notion of the present may be retained.

Keywords Special relativity • Presentism • Eternalism • Simultaneity • Einstein • Verificationism • Point presentism • Cone presentism

3.1 Introduction

It is often argued that the advent of special relativity has radically changed the concept of simultaneity and hence of time. Two opposed theses about the nature of this change have been intensely discussed in the philosophical literature.

First, it is widely argued that the fundamental difference between Newtonian physics and special relativity lies in the fact that simultaneity is no longer an absolute relation, i.e., the same for all observers, but rather depends upon a choice of an inertial frame. If we assume that two distant events e_1 and e_2 are simultaneous for me, and that one other observer is in constant motion relative to me, then e_1 and e_2 are ***not*** simultaneous for him. This property is known as the *relativity of simultaneity*.

Pursuing this astonishing consequence further, it is then argued that special relativity resolves the traditional dispute over the ontological status of the *Now*. Briefly,

Y. Cohen (✉)
Bar-Ilan University, Ramat Gan 5290002, Israel
e-mail: yehielc@gmail.com

Y. Dolev, M. Roubach (eds.), *Cosmological and Psychological Time*, Boston Studies in the Philosophy and History of Science 285,
DOI 10.1007/978-3-319-22590-6_3

it is said to refute one metaphysical view, the one known as *presentism*, according to which only present events are real, and to confirm a different metaphysical view called *eternalism*, according to which all events, whether past, present or future, are equally real. It is also said to prove that time passage, i.e., the becoming present and then past of future events, is not an objective feature of reality but just an illusion.

According to the opposite thesis, known as the *conventionality of simultaneity*, in a relativistic setting simultaneity within any inertial frame is merely conventional, i.e., it does not correspond to any objective constituent of physical reality. Hence, the relativity of simultaneity has no bearing on the problem of the Now.

My main purpose in this paper is to prove, by interrogating the logical relations between the two theses in light of the original metaphysical assumptions underlying special relativity, that presentism cannot be refuted by special relativity. To be more precise, I shall argue that the conventionality thesis is wrong, and also that the relativity of simultaneity is not an essential feature of special relativity and hence the pre-relativistic notion of an absolute present may be retained.

In Sect. 3.2 I first present the argument from special relativity against presentism and in Sect. 3.3 I discuss the bearing of the conventionality of simultaneity on this argument. In Sects. 3.4 and 3.5 I discuss two well-known attempts, due to Stein (1968) and Godfrey-Smith (1979), to fit presentism into a relativistic setting. Finally, in Sect. 3.6 I offer my own resolution.

3.2 The Alleged Refutation of Presentism by Special Relativity

The argument from special relativity was first introduced by Putnam (1967) though it was anticipated by Gödel (1949). Since then the details of it have been widely discussed, but in order to ensure a safe foothold for my own case I shall nevertheless go back to it.

The argument is quite straightforward. It is based upon two principal premises: presentism, i.e., the ontological principle that "all (and only) things that exist now are real" (1967, p. 240), and then what Putnam calls the "*There are no Privileged Observers*" principle, through which he proposes a partial explication of the notion of *reality*:

> If it is the case that all and only the things that stand in a certain relation R to me-now are real, and you-now are also real, then it is also the case that all and only the things that stand in the relation R to you-now are real. (*ibid*, p. 241)

This demands a closer look. The principle is often taken to mean that R is required to be transitive, yet it amounts to much more. As pointed out by Stein (1968, p. 19), the principle is to be reformulated as saying that "If yRz, then xRy *if and only if* xRz", where z and y mean "me-now" and "you-now", respectively. Taking note of the "*if and only if*" connective, we see that R is also required to be symmetric. And as R is obviously reflexive, the "No Privileged Observers" principle simply requires R to be an equivalence relation, i.e., the notion of reality which Putnam implicitly adheres to here is that ***reality*** is not dependent on any particular observer but is unitary. We shall revert to this important result later.

The argument then proceeds as follows. Consider an observer A at rest at a space-time point O_1, and suppose that another observer B, moving along the x^1-axis in constant relative motion, coincides with him at that point (Fig. 3.1). Let e_1 denote the event of their meeting. Next, we bring in special relativity. According to its most outstanding contentions, if events e_2 and e_1 are simultaneous for B then they are not simultaneous for A. As shown above in Fig. 3.1, e_1 chronologically precedes e_2, i.e., e_2 lies in A's future. However – and here comes the crux of the argument – B is obviously real for A at the point of their meeting, and e_2 is present for B at O_1, and hence real for him, then, employing the transitivity of the 'reality for' relation, A is compelled to admit that e_2 is real for him too!

Let us go further. Consider another future event of A at O_1,e_3, time-like separated from e_1, i.e., lies inside its light cone. According to special relativity, e_1 absolutely precedes e_3 and hence the above reasoning will not do. To prove that e_3 is real for A at O_1 we need to deploy a third observer. Suppose an observer C is at rest at a space-time point O_2, the location of e_2, such that e_3 is present for C at O_2. This, as depicted in Fig. 3.1, is a physically possible situation. Now, e_3 is obviously real for C at O_2, and, from our previous result, e_2, i.e., C at O_2, is real for B at O_1. Hence, by applying the transitivity of the 'reality for' relation once again, we see that B at O_1, and hence A at O_1, must affirm the reality for him of e_3 as well.

In general, any event in Minkowski space-time can be proved real for A at O_1 by simply adjusting the velocities of B and C. Hence, we are forced to renounce presentism and declare that "all future things are real, and likewise all past things are real, even though they do not now exist" (1967, p. 246).

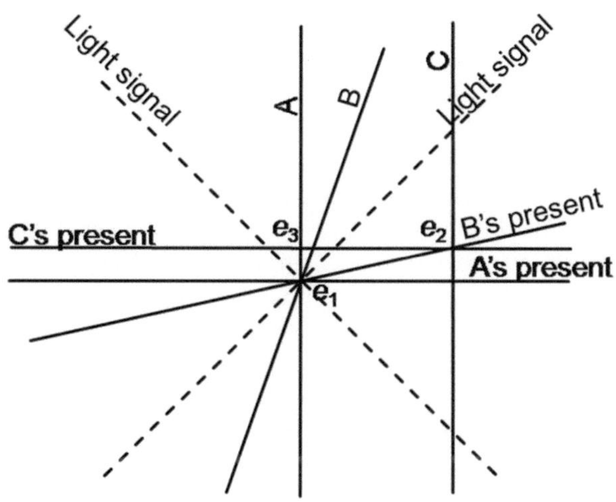

Fig. 3.1 Putnam's argument

3.3 Conventionality of Simultaneity

Many philosophers have taken Putnam's argument to have settled the debate on the side of eternalism. Nevertheless, the argument is not as conclusive as it seems at first.

Consider a simple solution of Sklar's (1981, p. 130) in refutation of Putnam's argument. Sklar argues that if we take presentism seriously, then adopting a frame-dependent notion of the present in a relativistic setting forces us to give up the idea of a single reality and to adopt a corresponding frame-dependent notion of reality. As a result, the 'reality for' relation is required to be neither symmetric nor transitive and therefore Putnam's purported conclusion, which crucially hinges on the transitivity of the 'reality for' relation, does not follow.

Ultimately Sklar does not endorse this relation (henceforth – $R_{relative}$). By drawing on Einstein's claim in his original special relativity paper that the simultaneity of spatially separated events cannot be empirically determined and hence must be established "*by definition*" (1905, p. 40), he argues:

> If we now associate the real (for an observer) with the simultaneous for him, we must, accepting the conventionality of simultaneity, accept as well a conventionalist theory of 'reality for'. It is then merely a matter of arbitrary stipulation that one distant event rather than another is taken as real for an observer. Now there is nothing inconsistent or otherwise formally objectionable about such a relativized notion of 'reality for', but it does seem to take the metaphysical heart out of the old claim that the present had genuine reality and the past and future lacked it. For what counts as the present is only a matter of arbitrary choice, and so then is what is taken as real. (1981, pp. 135–136)

Here Sklar calls attention to another profound change in the notion of simultaneity in special relativity, whose significance to the problem of the Now, it would seem, has not been sufficiently stressed. Indeed, as we shall see later, I believe that the difficulty posed by Sklar regarding $R_{relative}$ applies to Putnam, who also argues speciously that the fact that an event e_2 stands in a simultaneity relation with an event e_1 entails that it is also real for it.

But let us look first at the conventionality of simultaneity. Following Einstein, the classical argument for this thesis is due to Reichenbach (1958, pp. 123–135). Consider two identically constructed clocks U_a and U_b located at two distant points A and B in a given inertial frame, respectively, and let us suppose that a signal is propogated from A towards B where it is instantly reflected back (Fig. 3.2). Let events e_1 and e_3 denote the emission and subsequent return of the signal to A at times t_1 and t3, as indicated by U_a, and let event e_2 denote the reflection of it at B at time t_2, as indicated by U_b. Our problem is to synchronize U_b with U_a, i.e., to define t_2 in terms of t_1 and t_3.

If arbitrarily fast signals were possible, argues Reichenbach, the synchronization could be easily performed. e_2 clearly occurs after e_1 and before e_3, i.e., $t_1 < t_2 < t_3$. As the signal's velocity increases, the interval $t_3 - t_1$ is shortened and the ambiguity of t_2 diminishes, allowing us in the long run to extrapolate the value of t_2 at the limit,

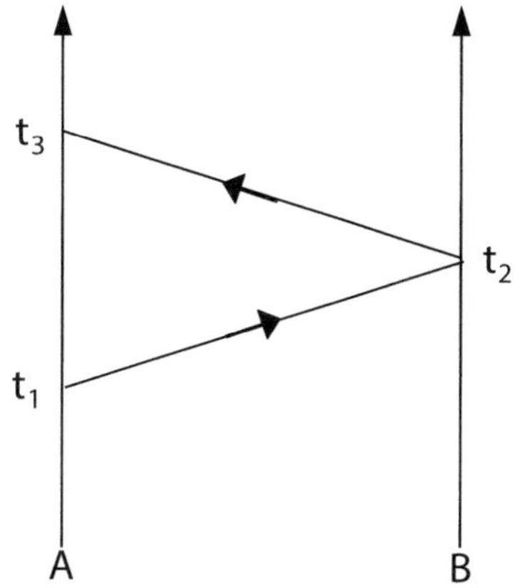

Fig. 3.2 Signal synchrony method

thus establishing a unique synchrony between U_a and U_b. Unfortunately, he argues, in a relativistic setting the speed of light is the upper limit of the velocity of signals and it is empirically determined to be finite. Hence, the ambiguity of t_2 cannot be dissolved.

However, consider now the second fundamental postulate of special relativity – *the principle of the constancy of the velocity of light:*

> Any ray of light moves in the "stationary" system of co-ordinates with the determined velocity c, whether the ray be emitted by a stationary or by a moving body. (1905, p. 41)

Hence, if *d* is the distance between the points then, to synchronize the clocks, we simply need to set U_b such that $t_2 = t_1 + d / c$. This reasoning, argues Reichenbach, will not work either. The velocity *c* in above postulate pertains only to the average round-trip velocity of light, empirically given by $c = 2d / (t_3 - t_1)$. The one-way velocity, in contrast, say along *AB* given by $c' = d / (t_2 - t_1)$, cannot be measured by means of a single clock $U_{a,}$ but requires prior synchrony between U_a and U_b. Hence, the proposed reasoning leads to a vicious circle.

Next, Reichenbach employs the verifiability theory of meaning, according to which, roughly speaking, the meaning of a statement is reducible to the conditions of its verification. As pointed out by Sklar (1981), (and many others) Einstein's motivation for introducing special relativity had to do with the physical operations for ascertaining, e.g., that two distant events are simultaneous – that is verificationism. Our inability to empirically dissolve the t_2 ambiguity, he argues, entails that simultaneity between spatially separated events is not a matter-of-fact, but must be established by convention. Any given event in the open interval (t_1, t_3) at A can be defined as simultaneous with e_2 at B, or to put in Reichenbach's famous ε-notation (1958, p. 127),

$$t_2 = t_1 + \varepsilon(t_3 - t_1) \text{ where} : 0 < \varepsilon < 1 \tag{3.1}$$

Einstein's definition, known as *standard synchrony*, sets ε as 1/2, thus entailing that light propagates isotropically, i.e., at the same one-way velocity in both directions.

Of course, the failure of the above signal synchrony method does not preclude other alternatives. And indeed, many methods, most notably the slow clock transport method (Ellis and Bowman 1967), have been introduced to show otherwise. Nevertheless, Winnie (1970, pp. 231–237) and Janis (1983, pp. 101–105) proved all such methods to be physically impossible in principle.

Let us now revert to Putnam's argument. Putnam argues, as shown above, that if B is real for A at O_1 and e_2 is present, and hence real, for B at O_1 then A at O_1 is forced to declare that e_2 is real for him as well, although it is in his future. However, I submit, given the conventionality of simultaneity, this reasoning is altogether misguided. Indeed, e_2 stands in a simultaneity relation with e_1 for B. However, since simultaneity between spatially separated events is merely conventional and ***not*** an objective constituent of reality, the above assertion is devoid of physical import. Hence, precisely as Sklar reasons in the quotation above, we cannot conclude that e_2 is real for B at O_1 and consequently for A at O_1. Thus, Putnam's alleged eternalism does not follow.

Surprisingly enough, eternalism does not follow even from Putnam's own premises. In accordance with the "No Privileged Observers" principle, Putnam set a constraint on the 'reality for' relation R. He argues that "R must be restricted to physical relations that are supposed to be independent of the choice of a coordinate system" (1967, p. 241), i.e., it must be definable in terms of the intrinsic geometrical elements and relations of Minkowski space-time. These correspond to events, light cones and inertial world lines, which are assumed to be empirically accessible to us by direct observation. However, standard simultaneity does not meet this constraint.[1] Hence, Putnam's use of it in his argument is entirely unwarranted.

3.4 Point Presentism[2]

Nevertheless special relativity still poses a serious threat to presentism. The "No Privileged Observers" principle requires the 'reality for' relation to be an equivalence relation and to be definable in terms of the intrinsic geometric structure of space-time. But if presentism is presupposed, can this principle be defined in Minkowskian space-time terms, so that it will not render ***all*** events real?

One well-known proposal is due to Stein (1968). He argues that "in Einstein-Minkowski space-time *an event's present is constituted by itself alone*" (*ibid*, p. 15), i.e., the 'reality for' relation is to be identified with the identity-relation.

[1] An ingenious attempt has been made in Malament (1977) to show that standard simultaneity is actually uniquely definable in terms of the intrinsic geometric structure of Minkowski space-time. Nevertheless, his result is still in considerable dispute. For more details, see Janis (2010).

[2] Here and in the next section I adopt Hinchliff's terminology (Hinchliff 2000).

Technically speaking, this relation (henceforth – R_{point}) satisfies the above requirement; however, some philosophers have argued that metaphysically speaking it is highly inadequate. First, it has been argued that R_{point} entails an untenable form of solipsism (see for instance Saunders (2002, p. 286), Callender (2000, pp. 592–593)). A single point – the here-now – cannot be all that is real for an observer. Second, why should an observer privilege his own here-now point as the sole constitute of reality over other observers' here-now points?[3]

Putnam (1967, p. 246) offers another objection to R_{point}. Consider an observer at rest at point A in a given inertial frame, and suppose that a light signal is propagated towards him from a distant point B. Let t denote the time of their meeting at A. The signal emission obviously never qualifies as present for A. However, after t it lies inside his backward light cone, so that it absolutely precedes any given event after that time. Hence, for these events the signal's emission becomes past without ever being present!

Stein's attempt to dispose of this alleged bizarre consequence is essentially similar to my response above to Putnam's argument from special relativity:

> In this theory, …, the present tense can *never* be applied correctly to "foreign" objects. This is at bottom a consequence (and a fairly obvious one) of our adopting relativistically invariant language – since, as we know, there is no relativistically invariant notion of [spatially extended] simultaneity. The appearance of paradox only confirms that the space-time of Einstein and Minkowski is quite different from pre-relativistic space-time. (1968, p. 15)

This is also the kernel of Stein's argument for rejecting Putnam's argument against presentism and advancing R_{point}.

But, now, I wish to reconsider my previous assertion on the conventionality of simultaneity, and argue as follows: Although Einstein's standard simultaneity ($\varepsilon = 1/2$) is not dictated by the facts, if one assumes a verifiability theory of meaning, then simultaneity between spatially separated events (not necessarily standard synchrony) in a relativistic setting ***is*** after all a matter-of-fact. Consequently, again, R_{point} cannot be sustained.

To see this, let us revert to the signal synchrony method, depicted in Fig. 3.2. The light signal is sent from A at t_1 and returns at t_3. The question arises: What happens to it while it is spatially separated from A in the open interval (t_1, t_3)? Certainly, it does not cease to exist according to special relativity, and as illustrated by the line denoting the light signal in Fig. 3.2, its world line is understood to be spatiotemporally continuous. Hence, for any given event e in (t_1, t_3) the light signal constitutes at least part of its reality!

This conclusion has a far-reaching impact on the conventionality of simultaneity and the problem of the Now. However, to achieve expository clarity, I shall postpone its full discussion to Sect. 3.6.

[3]This argument is due to Putnam (1967, p. 246) although he did not state it in terms of reality of points.

3.5 Cone Presentism

I now turn to another alternative relation, proposed by Godfrey-Smith (1979), that can be identified with the 'reality for' relation, which does not lead to eternalism. Godfrey-Smith argues that in a relativistic setting "the present be identified with the class of events which are 'seen now' by an observer" (*ibid*, p. 240), i.e., the reality of any given event is constituted by the set of events which lie on its backward light cone.

Interestingly enough, this relation (henceforth – R_{bcone}) was first considered by Einstein himself in his original paper as a possible definition of simultaneity:

> We might, of course, content ourselves with time values determined by an observer stationed together with the watch at the origin of the co-ordinates, and co-ordinating the corresponding positions of the hands with light signals, given out by every event to be timed, and reaching him through empty space. (1905, p. 39)

but eventually, as is well-known, he chose standard simultaneity for practical considerations. Godfrey-Smith, in contrast, argues that there is an essential difference between the two definitions. R_{bcone}, unlike standard simultaneity, is definable in terms of the intrinsic geometric structure of Minkowski space-time and hence only R_{bcone} qualifies as the 'reality for' relation.

Another interesting virtue of R_{bcone} was pointed out by Clifton and Hogarth (1995, p. 364), who do not endorse it themsleves. Suppose an observer could – although obviously he physically cannot – travel at the speed of light then he would lack any psychological sense of time passage. This follows directly from the fact that a clock in relative motion slows down as a consequence of its motion. If, for example, Δt is a time interval as determined by a clock at rest and v is the velocity of the moving clock then according to the Lorentz transformations its corresponding time interval is given by $\Delta t' = \Delta t\sqrt{1-(v^2/c^2)}$. For the velocity $v = c$ we have $\Delta t' = 0$, i.e., for the moving observer any two events are perceived as occurring simultaneously.

Despite these advantages, I think that R_{bcone} is untenable. First and most importantly, as shown above regarding the continuous existence of the light ray,.for any given event some events that are spatially separated from it are real. Hence, although we are no longer alone, R_{bcone} is obviously deficient as a description of reality.

Second, R_{bcone} is neither a symmetric nor a transitive relation whereas the 'reality for' relation is required to be an equivalence relation. In Fig. 3.2, for example, e_3 at A stands in a relation of R_{bcone} to e_2 at B, but e_2 does stand in the R_{bcone} relation to e_3. Moreover, e_3 stands in a relation of R_{bcone} to e_2 and e_2 stands in relation R_{bcone} to e_1 at A. But lying on the same world line of a material object, i.e., clock U_a, e_3 obviously does not stand in a R_{bcone} relation to e_1.

The proponent of R_{bcone} could easily avoid this objection by denying the assumption that reality is unitary and argue that it conforms to a privileged space-time point – the apex of the backward light cone. A similar solution, as shown above, was suggested by Sklar in the form of his $R_{relative}$. But I think there is a crucial difference between the two cases. Sklar argues that, following the relativity of simultaneity, the presentist must adopt a corresponding relativized notion of reality. Yet, no compelling reason has been provided in the case of R_{bcone} and hence omitting symmetry and transitivity seems unjustified.

The last objection to be mentioned is due to Savitt (2000, p. 567): If any event is present for a given event *e* then, he argues, it cannot chronologically precede or succeed it. In particular, it obviously cannot be either timelike or lightlike separated from *e*. However, this seemingly minimal achronality requirement is violated by R_{bcone} since all the events which stand in relation of R_{bcone} to *e* are, by definition, lightlike separated from it.

3.6 Spatially Extended Presentism

As I argued in the last two sections, the main weakness of both R_{point} and R_{bcone}, is that reality is necessarily constituted by spacelike-separated events. Hence, it appears that the only way to retain presentism is to retain the pre-relativistic notion of an absolute spatially extended present.

This solution is actually implied by special relativity. To see this, let us re-examine the link between the conventionality of simultaneity and the relativity of simultaneity. Consider the following passage of Savitt's:

> [The controversial thesis that simultaneity is conventional] is not to be confused with the thesis that simultaneity is relative. The latter is non-controversially a feature of special relativity. (2000, p. 571 fn. 15)

Although, this sort of claim has often been repeated, it is obviously wrong. First, as shown above in Sect. 3.3, given the conventionality thesis, the fact that adopting the same definition of simultaneity, $\varepsilon = 1/2$, in all inertial frames yields different simultaneity relations is devoid of physical significance.

Second, as pointed out by Reichenbach (1958, p. 146) and later on by Grünbaum (1963, pp. 359–368), we can define an absolute spatially extended simultaneity relation even in a relativistic setting by employing an appropriate distribution of ε values. This result follows directly from the interesting fact that simultaneity defined by $\varepsilon = 1/2$ in a given inertial frame corresponds to $\varepsilon \neq 1/2$ in another. Hence, to define an absolute simultaneity relation, we simply need to fix $\varepsilon = 1/2$ in an arbitrary inertial frame and then adjust the ε values in the remaining inertial frames such that they will lead to the same result.

Consider, for example, an inertial frame K and suppose that another inertial frame K′ travels in a relative motion along its x^1-axis. Let the ε values in K and K′ be ε_1 and ε_2, respectively. According to transformations formulated by Winnie (1970, pp. 229–237), known as ε-Lorentz transformations, in which ε appears as a free parameter (whereas the ordinary Lorentz transformations implicitly take ε as 1/2), the time coordinate of K′ is given by

$$t' = \frac{\left\{\left(\frac{2v_{\varepsilon_1}}{c}\right)(1-\varepsilon_1-\varepsilon_2)+1\right\}t - x\left\{\frac{2c(\varepsilon_1-\varepsilon_2)+4v_{\varepsilon_1}(\varepsilon_1)(1-\varepsilon_1)}{c^2}\right\}}{\sqrt{\frac{\left(c-v_{\varepsilon_1}(2\varepsilon_1-1)\right)^2-v^2_{\varepsilon_1}}{c^2}}} \tag{3.2}$$

Now, if we take $\varepsilon_1 = 1/2$ and, for simplicity's sake, we assume $t = 0$ and $x \neq 0$ then to establish an absolute simultaneity relation we must choose

$$\varepsilon_2 = \frac{1}{2}\left(1+\frac{v_{1/2}}{c}\right). \tag{3.3}$$

Next, we bring in the fact illustrated in Fig. 3.2, that for any given event in (t_1, t_3) at A the light signal constitutes at least part of its reality. Coupled with the empirical fact that light travels at a finite speed, so that it cannot occupy two distinct points along its path, it entails that e_2 at B is real for only a single event in (t_1, t_3). That is, having identified the real with the present, the verifiability theory of meaning actually suggests a weak reading of the conventionality thesis, according to which all choices of ε, where $0 < \varepsilon < 1$, are empirically equivalent; but only a single value, I say, and hence a single distribution of ε values in all inertial frames, corresponds to actual reality.

The implications for the problem of the Now are quite straightforward. First, as shown above, the relativity of simultaneity is not an essential feature of special relativity but depends upon a distribution of ε values in all inertial frames. Interestingly, some distributions yield an absolute spatially extended simultaneity relation and it is physically possible that one of them should coincide with reality. Hence, presentism is not refuted by special relativity!

3.7 Conclusion

Putnam concludes his paper with the following passage:

> I conclude that the problem of reality and determinateness of future events is now solved. Moreover, it is solved by physics and not by philosophy. …Indeed, I do not believe that there are any longer philosophical problems about Time… (1967, p. 247)

This assertion I believe to be wrong. For as is well-known, Einstein's philosophical motivation for advancing special relativity is essentially verificationist. However, as shown above in Sect. 3.6, given the verifiability theory of meaning, the pre-relativistic notion of an absolute simultaneity relation is not only compatible but can be implied by special relativity. Hence, presentism, in its original form, can be retained.

Nevertheless, I believe that the dispute over the importance of special relativity for the problem of the Now is far from being settled. The verifiability theory of meaning with its distinction between matters-of-fact and conventions, has long been considered untenable. Unfortunately, in the voluminous literature on special relativity and, in particular, on the conventionality of simultaneity, this fact has been often neglected. And so a fresh analysis of the dispute within the framework of recent philosophy is required. Until then, the problem of the Now remains an open question.

References

Callender, C. 2000. Shedding light on time. *Philosophy of Science* 67: 587–599.

Clifton, R., and M. Hogarth. 1995. The definability of objective becoming in Minkowski space-time. *Synthese* 103: 355–387.

Einstein, A. 1905/1952. On the electrodynamics of moving bodies. In *The principle of relativity*, 35–65. New York: Dover Publications.

Ellis, B., and P. Bowman. 1967. Conventionality in distant simultaneity. *Philosophy of Science* 34: 116–136.

Gödel, K. 1949. A remark about the relationship between relativity and idealistic philosophy. In *Albert-Einstein: Philosopher-Scientist*, ed. P. Schilpp, 557–62. La Salle: Open Court.

Godfrey-Smith, W. 1979. Special relativity and the present. *Philosophical Studies* 36: 233–244.

Grünbaum, A. 1963. *Philosophical problems of space and time*. New York: Knopf.

Hinchliff, M. 2000. A defense of presentism in a relativistic setting. *Philosophy of Science* 67: 575–586.

Janis, A. 1983. Simultaneity and conventionality. In *Physics, philosophy and psychoanalysis*, ed. R. Cohen and L. Laudan, 101–110. Dordrecht: Reidel. First published Mon Aug 31, 1998; substantive revision Wed Jul 16, 2014.

Janis, A. 2010. Conventionality of simultaneity. *Online Stanford Encyclopedia of Philosophy.*

Malament, D. 1977. Causal theories of time and the conventionality of simultaneity. *Noûs* 11: 293–300.

Putnam, H. 1967. Time and physical geometry. *Journal of Philosophy* 64: 240–247.

Reichenbach, H. 1958. *The philosophy of space & time*. New York: Dover Publications.

Saunders, S. 2002. How relativity contradicts presentism. In *Time, reality & experience*, ed. C. Callender, 277–292. Cambridge, MA: Cambridge University Press.

Savitt, S. 2000. There's no time like the present (in Minkowski space-time). *Philosophy of Science* 67: 563–574.

Sklar, L. 1981. Time, reality and relativity. In *Reduction, time and reality*, ed. R. Healy, 129–142. New York: Cambridge University Press.

Stein, H. 1968. On Einstein-Minkowski space-time. *The Journal of Philosophy* 65: 5–23.

Winnie, J. 1970. Special relativity without one-way velocity assumptions: Part II. *Philosophy of Science* 37: 223–238.

Chapter 4
Einstein's Bergson Problem: Communication, Consensus and Good Science

Jimena Canales

Abstract Does a privileged frame of reference exist? Part of Einstein's success consisted in eliminating Bergson's objections to relativity theory, which were consonant with those of the most important scientists who had worked on the topic: Henri Poincaré, Hendrik Lorentz and Albert A. Michelson. In the early decades of the century, Bergson's fame, prestige and influence surpassed that of the physicist. Once considered as one of the most renowned intellectuals of his era and an authority on the nature of time, *The Stanford Encyclopedia of Philosophy* (2010) does not even include him under the entry of "time." How was it possible to write off from history a figure that was once so prominent? Through an analysis of behind-the-scenes of science correspondence, this article traces the ascendance of Einstein's views of time at the expense of Bergson's.

Keywords Einstein • Bergson • Time • Philosophy • Physics • Twin paradox • Frame of reference • Relativity • *Duration and Simultaneity*

4.1 Introduction

Imagine the famous twins of relativity theory talking to each other through some form of long-distance communication. Peter says to Paul: "The moment you separated from me … your time swelled, your clocks disagreed" with mine. Just imagine what "Paul would reply." That everything was normal for him (Paul) and that it was Peter's system that had gone awry. Who was right? According to Einstein, both were right. If they were as rational as scientists, they would eventually come to that conclusion.

Einstein's theory of relativity has been widely confirmed by a wide array of tests and experiments. Scientists typically refer to the "three classic tests" of general

J. Canales (✉)
University of Illinois, 810 S. Wright St, Urbana, IL 61801, USA
e-mail: jcanales@illinois.edu

Y. Dolev, M. Roubach (eds.), *Cosmological and Psychological Time*, Boston Studies in the Philosophy and History of Science 285,
DOI 10.1007/978-3-319-22590-6_4

relativity as the perihelion of mercury, the red shift, and the gravitational bending of starlight. The special theory of relativity was strikingly confirmed when in 1972 scientists transported an atomic clock eastward around the world and compared it with one transported westward. The far-east traveler lost 59 ns, while the one transported westward gained 273 ns.[1] Other experiments with cosmic-ray muons (particles that enter into the Earth's atmosphere from outer space) showed that their lifespan before decaying was noticeably increased. Scientists interpreted the particles' prolonged life as due to time dilation effects arising from the relativity effects of traveling at speeds close to that of light.

Consider this additional experiment, brought up by the philosopher Henri Bergson, only a few months before Einstein was awarded the Nobel Prize. What would happen if one of the twins was so dominant, that he would not accept that his travelling sibling was seeing just the opposite he was, but who would stubbornly insist that something strange was going on in his brother's system. Everything was really normal for *him* (and only him) and it was the other twin who was going through strange dilation changes regardless of how he saw things. This dominant twin might want to force the other one to use Lorentz's equations to *correct* his perceptions and return to real life. "Here are the correction formulas that will permit you to come back to reality." Use them! he might insist. This would be a great victory for the dominant twin: "I can go on as if none of my lengths have shrunk, as if my time has not dilated, as if my clocks where working fine."[2] The less dominant twin, in contrast, would find himself in deep trouble: "I would have to completely redo the science of electromagnetism, which you have so laboriously constructed: I would have to modify [the results of] my equations, once I established them, every time I change velocities."[3] Thankfully, there was an evident solution that would prevent one of them from going through this hardship, but it would entail a compromise.

Does it matter, for science, to consider what the twins might say to each other? Would science have to take into consideration a twin so stubborn that he would not accept that the changes his other was going through as symmetrical and reciprocal? Not at all. After all, the hypothetical disagreement among the twins could be solved by recording devices that would attest that both of them were going through exactly reciprocal processes. When Paul Langevin first imagined the "voyage à boulet"—later baptized as the "twin paradox"—he imagined ways in which the two clocks could be compared while remaining at a distance from each other: "It is fun to imagine how our explorer and the planet Earth would see each other mutually live, if they could stay in constant communication by light signals or by wireless telegraphy, during separation, and thus understand how the asymmetry between two mea-

[1] J.C. Hafele and Richard E. Keating, "Around-the-World Atomic Clocks: Predicted Relativistic Gains," *Science*, no. 177 (1972).

[2] Henri Bergson, *Durée et simultanéité: á propos de la théorie d'Einstein*, ed. Élie During, 4th ed. (Paris: Quadrige/Presses Universitaires de France, 2009b). 17.

[3] Ibid., 18.

sures of time is possible."[4] These speculations made sense in light of the surge in the development of wireless technology from 1905 (the date of Einstein's paper) to 1911 (the date of Langevin's).[5]

Does it matter philosophically? The answer can be more complicated. *If only* we were just like instruments the debates pertaining to time dilation in relativity theory would be perfectly solved. Yet solving this question by reference to what instruments measure overlooks the broader question of the role of scientific instruments in the world and their relation to living consciousness. We are not just like clocks or recording instruments, argued Bergson. If we were, we would be living in a world without consciousness, where we might as well "say goodbye to the theory of relativity" in its entirety.[6]

Bergson's objections hurt Einstein enough that the Nobel Prize Committee decided not to award him the prize for relativity theory. The president of the committee explained that although "most discussion centers on his theory of relativity" it did not merit the prize. Why not? Reasons were surely varied and complex, but the culprit mentioned that evening was clear: "It will be no secret that the famous philosopher Bergson in Paris has challenged this theory." Bergson had shown that relativity "pertains to epistemology" rather than to physics—as it "has therefore been the subject of lively debate in philosophical circles."[7]

How did the debate between Einstein and Bergson unravel? How were they able to come to an agreement? As it turns out, an agreement was never reached. Bergson's objections were successfully labeled as essentially mistaken. Einstein himself led the initiative against the famous philosopher.

Might does not make right, especially not in science. "One of the strongest, if still unwritten, rules of scientific life is the prohibition of appeals to heads of state or to the populace at large in matters scientific," explained the historian and philosopher Thomas Kuhn in his *Structure of Scientific Revolutions*.[8] Science, truth and democracy, can and should go together. Our very idea of scientific knowledge as essential to modern civilization and superior to other ways of knowing relies on this notion. Of course, many scholars and scientists agree that power relations affect science, but only once they are tamed by institutions and professional codes of conduct designed to foster civil, free discourse and guarantee ethical standards. Truth and knowledge may still go hand-in-hand with power, but not in an overt or forced way. "Mob rule" has no say in science.[9]

---

[4] Paul Langevin, "L'evolution de l'espace et du temps," *Scientia* 10 (1911).

[5] Jimena Canales, "The Media of Relativity: Einstein and Communications Technologies," *Technology and Culture* (2015).

[6] "Appendix III" in Bergson, *Durée et simultanéité: á propos de la théorie d'Einstein*: 207 n. 201.

[7] Svante Arrhenius, 10 December 1922 in *Nobel Lectures in Physics (1901–1921)*. (Singapore: World Scientific, 1998), 479. Italics mine.

[8] Thomas S. Kuhn, *The Structure of Scientific Revolutions*, 2nd ed. (Chicago: University of Chicago Press, 1970). 168.

[9] Imre Lakatos, "Falsification and the Methodology of Scientific Research Programmes," in *Criticism and the Growth of Knowledge*, ed. Imre Lakatos and Alan Musgrave, *Studies in Logic*

Over the last two decades we have learned much more about how scientific controversies are actually settled. Ethnographic, anthropological, sociological and historical studies of science increasingly reveal a large gap between the actual practice of science and normative views about what science "should be." Discussions about how controversies *ought* to be settled needs to be rethought in terms of how they *are* settled. Recent literature on controversies has focused on the role of institutions in vetting selected experts to evaluate knowledge-claims, the use of history for establishing novelty, strategies of social marginalization, public and dramatic staging, pedagogical and textbook initiatives, the role of unconscious bias, the recruitment of allies (human and non-human), and techniques for extending results obtained in laboratory conditions outwards. The contested facts in question and of the instruments and experiments connected to them are key as well. Certain facts and experiments are easier to defend and reproduce than others. Some travel easily whereas others are hard to move. In some notorious controversies, disputes do not even hinge on matters of fact, but continue even after agreement about them has been reached. Facts do not speak for themselves.

One of Einstein's most successful victories consisted in eliminating Bergson's objections to relativity theory. In the early decades of the century, Bergson's fame, prestige and influence surpassed that of the physicist—who is, in contrast, so well known today. Once considered as one of the most renowned intellectuals of his era and an authority on the nature of time, *The Stanford Encyclopedia of Philosophy* (2010) does not even include him under the entry of "time."[10]

How was it possible to write off from history a figure that was once so prominent? Einstein met Bergson for the first time during his trip to Paris on April 6, 1922. That day the philosopher advanced some arguments against Einstein's interpretation of relativity theory. He was about to publish a book on it, titled *Duration and Simultaneity*, which would appear later that year. Bergson spoke for about half an hour. "We are more einsteinian than you, Monsieur Einstein," he explained.[11] The physicist responded in less than a minute—including in his answer one damning and frequently cited sentence: "*Il n'y a donc pas un temps des philosophes.*"[12]

The meeting between the two men caused such a stir in intellectual circles, that both men came to represent opposing views about time in the work of intellectuals as diverse as Martin Heidegger, Alfred N. Whitehead, George Herbert Mead, Gaston Bachelard, Maurice Merleau-Ponty and more recently Gilles Deleuze and Bruno Latour. While Einstein's notion of time would be frequently referred to as "objec-

---

and the Foundations of Mathematics (Cambridge, UK: University Press, 1970). For the continuing fear of "mob rule" in science see Bruno Latour, " Do You Believe in Reality? News from the Trenches of the Science Wars," in *Philosophy of Technology: The Technological Condition*, ed. Robert C. Scharff and Val Dusek (Malden, MA: Blackwell Publishers, 2003).

[10] Ned Markosian, "Time," in *Stanford Encyclopedia of Philosophy*, ed. Edward N. Zalta (Winter 2010).

[11] "La Théorie de la relativité: séance du 6 avril 1922," *Bulletin de la Société française de philosophie* 22, no. 3 (1922). Re-phrased as "More einsteinian than Einstein," in Bergson, *Durée et simultanéité: á propos de la théorie d'Einstein*: 55.

[12] "La Théorie de la relativité: séance du 6 avril 1922," 364.

tive" and "physical" and the time of Bergson as "subjective" and "psychological," neither of these labels do justice to the contributions of each men.

4.2 Erased from Einstein's Life

Einstein's visit to Paris was covered in sensational detail by the French and German press.[13] Yet, for the most part, Einstein's biographers have ignored the role played by Bergson. The authoritative biographies of Philipp Frank, Ronald W. Clark, Albrecht Fölsing, Jürgen Neffe, Abraham Pais, Walter Isaacson, among others, either do not mention Bergson or discount his importance.[14] His absence in these retrospective accounts is especially notable compared to the importance of the episode during the period. Einstein had to go to great lengths to hide from journalists that day and to make sure that no one knew where he was staying.

For many years Bergson was portrayed as someone who did not *understand* the physics of relativity. "Bergson is mistaken," insist the physicists Alan Sokal and Jean Bricmont, adding that his "error is not a question of philosophy or interpretation, as is frequently thought; it bears on understanding the physical theory, and it enters, in the last analysis, in conflict with experience."[15] Sokal and Bricmont cited Bergson's remark that "once reentering [Earth], it [one clock] marks the same time

[13] Michel Biezunski, *Einstein à Paris: le temps n'est plus...* (Vincennes: Presses Universitaires de Vincennes, 1991).

[14] Philipp Frank focused on Einstein's physics lecture the day *before* at the Collège de France, and not the one the following day at the Société française de philosophie, when Bergson delivered his criticisms. Clark recalled how Einstein "was closely questioned" by French philosophers and does not mention Bergson in that context. In places where Fölsing deals with the relation between Einstein and contemporary philosophers, Bergson intervention is not mentioned (although he is mentioned in another context). Jürgen Neffe's biography (2005, German edition) contains no mention of Bergson. Abraham Pais mentions the meeting between Einstein and Bergson only to downplay any negative role he may have played in Einstein's life. Pais states that the special theory of relativity "caused confusion in philosophical circles, as witness, for example the little book on the subject by Henry Bergson written as late as 1922." He claims that "Einstein came to know, like, and respect Bergson. Of Bergson's philosophy he used to say, '*Gott verzeih ihm*', God forgive him." Pais mentions Bergson again in his account of "How Einstein Got the Nobel Prize" where he explicitly discounts the role of Bergson. He points out instead to the compilation of Bergson's collected works that excluded *Duration and Simultaneity* from the collection. Le Roy explained his decision not to republish *Duration and Simultaneity* in Le Roy, Lettre-Préface, 29 September 1953 in Henri Bergson, *Écrits et paroles*, ed. Rose-Marie Mossé-Bastide, Bibliothèque de philosophie contemporaine (Paris: Presses Universitaires de France, 1957). vii–viii. Some of these references were provided to me by an anonymous reviewer for Chicago University Press. Philipp Frank, *Einstein, His Life and Times*, trans. George Rosen (New York: A.A. Knopf, 1947). 196; Ronald William Clark, *Einstein: The Life and Times* (New York: World Pub. Co., 1971); Albrecht Fölsing, *Albert Einstein: A Biography* (New York: Viking, 1997); Jürgen Neffe, *Einstein: A Biography*, 1st American ed. (New York: Farrar, Straus, and Giroux, 2007); Abraham Pais, "*Subtle is the Lord...*": *The Science and the Life of Albert Einstein* (Oxford: Clarendon Press, 1982). 28, 163, 510.

[15] Alan D. Sokal and J. Bricmont, *Impostures intellectuelles* (Paris: O. Jacob, 1997).

as the other" as proof of his profound misunderstanding of relativity.[16] They were only repeating what had been said many times before them, starting with Einstein. Because of this particular claim, many readers insisted that Bergson "was not sufficiently conversant with the outlook and problems of mathematics and physics."[17] "These attempts [Bergson's] … have totally failed: science, on this issue, has passed purely and simply to become the order of the day."[18] By the 1960s Bergson's fate as somebody who simply did not understand science was sealed: "The best explanation for Bergson's impressive failure as a scientific theoretician is the same as that for his failure to succeed as a metaphysician: he was not sufficiently conversant with the outlook and problems of mathematical physics."[19] Even a writer in the *Annales Bergsoniens*—an ongoing series solely dedicated to his philosophy—stated that "Bergson could not understand him [Einstein]."[20]

The chemist and Nobel Prize winner Ilya Prigogine reviewed the collection of essays that included translated portions of the transcript of the April 6, 1922 meeting for the journal *Nature*. Prigogine was extremely critical of Bergson's "pathetic" work on relativity: "Bergson's struggle with the Lorentz transformation in *Duration and Simultaneity* is as pathetic as it completely misses the point."[21]

4.3 Contrast Between Einstein's Journal and Letters

Let us retrace our steps and go back to the moment after the meeting. Only a few months after their encounter in Paris, Einstein was asked once again what he thought of Bergson. He had "received the Bergson book and read part of it but have not yet been able to make up my mind about it finally," he explained to Lord Haldane.[22] Later that fall, he finally found time to examine it more carefully. Einstein took Bergson's book with him on the boat that would take him to Japan. The day the ship exited the harbor he started reading, and the next morning he jotted down some

[16] Alan D. Sokal and J. Bricmont, *Fashionable Nonsense: Postmodern Intellectuals' Abuse of Science* (New York: Picador, 1998). 176. They also attribute this error to Merleau-Ponty's understanding of relativity.

[17] Thomas Hanna, "Introduction," in *The Bergsonian Heritage*, ed. Thomas Hanna (New York: Columbia University Press, 1962b), 23.

[18] André George, *Les documents de la vie intellectuelle* (January 1931): 60. Cited in André Metz, "Bergson, Einstein et les relativistes," *Archives de philosophie* 22(1959): 378.

[19] Thomas Hanna, ed. *The Bergsonian Heritage* (New York: Columbia University Press, 1962a), 23. For a later text on this topic see Andrew C. Papanicolaou and P.A.Y. Gunter, *Bergson and Modern Thought: Towards a Unified Acience*, Models of Scientific Thought (Chur, Switzerland: Harwood Academic Publishers, 1987).

[20] Hervé Barreau, "Bergson et Einstein: à propos de Durée et simultanéité," *Les Études bergsoniennes*, no. 10 (1973): 167.

[21] Ilya Prigogine, "Evolution of Physics: Review of Bergson and the Evolution of Physics, edited and translated by of P.A.Y. Gunter," *Nature* 234(1971): 159.

[22] Einstein to Richard B. Haldane, 11 September 1922, Berlin.

quick notes about it in his travel journal. He acknowledged that Bergson had fully "grasped the substance relativity theory" and considered the philosopher's contribution as one that merely "objectivized" *psychological* aspects of time: "The philosophers constantly dance around the dichotomy: the psychologically real and physically real, and differ only in evaluations in this regard."[23]

Einstein felt the damaging effects of Bergson's critique a few months later. When he was awarded the Nobel Prize (in December 1922 for the previous year), it was not given for the theory that had made the physicist famous: relativity. Instead it was given "for his discovery of the law of the photoelectric effect"—an area of science that hardly jolted the public's imagination to the degree that relativity did. The reasons behind the decision to focus on work *other* than relativity were directly traced to Bergson's intervention in Paris.

The Bergson problem was not going away for Einstein. *Duration of Simultaneity* was a success, albeit a controversial one. An augmented revised version appeared the following year, containing three new appendixes aimed at responding to critics. In 1936, less than a decade and a half after it first appeared, an avid reader warned prospective buyers that they "might have difficulty in finding" a copy "as the last edition is exhausted."[24]

Bergson and Einstein met in other occasions after their encounter in Paris. Bergson was the president of the CIC, a branch of the League of Nations and Einstein was a member. The CIC was founded on the idea that if intellectuals set the example for peaceful cooperation, then the world might follow. The philosopher Isaac Benrubi, amongst others, decided to attend the CIC's meeting in Geneva (25 July 1924) after learning that both Einstein and Bergson would attend.[25]

For the meeting Bergson introduced Einstein in flattering terms, but during the meeting *break* their differences once again became evident. Benrubi approached Einstein to ask him what he thought of *Duration and Simultaneity*. Einstein offered his official response, that Bergson had not understood the *physics* of relativity and that he had made a mistake. Asked if he would continue the fight against Bergson, Einstein responded: "No, I do not intend to do that, unless Bergson himself pro-

---

[23] Einstein, travel diary to Japan, Palestine, Spain, [6 October 1922 to 12 March 1923]. See the entry for 9 October 1922: "Yesterday I looked into Bergson's book on relativity and time. Strange that time alone is problematic to him but not space. He strikes me as having more linguistic skill than psychological depth. He is not very scrupulous about the objective treatment of psychic factors. But he does seem to grasp the substance of relativity theory and doesn't set himself in opposition to it. The philosophers constantly dance around the dichotomy: the psychologically real and physically real, and differ only in evaluations in this regard. Either the former appears as a "mere individual experience" or the second as "mere construct of thought." Bergson belongs to the latter kind but objectifies in *his* way without noticing." For an account of Einstein's impressions of Bergson from his travel diary see Armin Hermann, *Einstein: Der Weltweise und sein Jahrhundert; eine Biographie* (München: Piper, 1996). 283.

[24] Pierre Lecomte du Noüy, *Biological Time* (London: Methuen and Co. Ltd., 1936). 127.

[25] Isaac Benrubi, *Souvenirs sur Henri Bergson* (Neuchâtel: Delachaux & Niestlé, 1942).

vokes a polemic. But that would not help anybody."[26] Was Einstein willing to let bygones be bygones?

The debate nonetheless continued to explode, especially after André Metz, an alumnus of the École polytechnique and army captain stationed at the Rhine reignited the polemic by publishing damaging articles in favor of Einstein and against Bergson in the *Revue de philosophie*.[27] Einstein enthusiastically endorsed Metz's work on relativity. It "responded to a real need," "was completely exact" and contained the "refutation of the inexact assertion of other authors."[28] To observers at the time, it was clear that Metz and Einstein had a lot in common—even personally. "The personal philosophy of Einstein is similar to that of Metz," explained the Catholic philosopher Jacques Maritain.[29]

Metz was a soldier who believed that confrontations "at times violent" were "necessary conditions" for "making history." This maxim was true, he explained for "all domains"—including science: "The *triumph* of people, ideas or theories seems to be a necessary condition for *the fight*, and of a bitter struggle, sometimes violent." Metz underlined how Einstein would only prevail if he fabricated controversies, fought hard and won. In a private letter he described his intentions clearly: "The names which remain in history are those of men who have fought, and that in all areas." He considered Einstien within this category: "Einstein himself, with his simple and benevolent disposition, owes his fame to the controversies his theory raises."[30]

In a private letter to Metz, Einstein framed Bergson's mistake in terms of physics. Metz did not hesitate to publish it in the prestigious *Revue de philosophie*. "It is regrettable that Bergson should be so thoroughly mistaken, and his error *is really of a purely physical nature*, apart from any disagreement between philosophical schools," explained Einstein. He spelled-out Bergson's "mistake" in detail: "Bergson forgets that the simultaneity … of two events which affect one and the same being is something absolute, independent of the system chosen."[31] Einstein, again, used

---

[26] "Es wird Gras darüber wachsen, und dann wird man mit mehr Objectivität darüber urteilen." Ibid., 107–108; Angelo Genovesi, "Henri Bergson: Lettere a Einstein," *Filosofia* 49, no. 1 (1998): 8–9; Rose-Marie Mossé-Bastide, *Bergson éducateur* (Paris: Presses Universitaires de France, 1955). 126.

[27] André Metz, "Le Temps d''Einstein et la philosophie: à propos de l'ouvrage de M. Bergson, Durée et simultanéité," *Revue de philosophie* 31 (1924a); André Metz, *La Relativité: Exposé dans formules des théories d'Einstein et réfutation des erreurs contenues dans les ouvrages les plus notoires* (Paris: Etienne Chiron, 1923). On Metz see Ramiro Ledesma, "Actualidad. Filosofía, Ciencia: Andre Metz y la Geometría Euclidiana," *La Gaceta Literaria* (1929); Metz, Bergson, Einstein et les relativistes."; Eva Telkes-Klein," Meyerson dans les milieux intellectuels français dans les années 1920," *Archives de Philosophie* 70, no. 3 (2007): 370.

[28] Einstein's comments were included in the second edition.

[29] Jacques Maritain, *Réflexions sur l'intelligence et sur sa vie propre* (Paris: Nouvelle Libraire Nationale, 1926 (1924)). 225 n. 221.

[30] Archives Centre Sèvres, 408/70, folder Metz, 12 January 1925. Cited in Telkes-Klein, "Meyerson dans les milieux intellectuels français dans les années 1920," 371.

[31] Einstein to Metz, 2 July 1924. Published after his reply, André Metz, "Replique a Bergson," *Revue de philosophie*, no. 31 (1924b). Republished in P.A.Y. Gunter, ed. *Bergson and the Evolution of Physics* (Knoxville: The University of Tennessee Press, 1969), 189–190.

the word "absolute" to show that thinking about relativity in terms of the *difference* between the two travelers was essentially misguided.

Almost three years after the meeting between Einstein and Bergson took place, a scientist and popular science writer from Barcelona could still expect readers in Spain to be "aware of Bergson's objections" to Einstein, and of the particular occasion when the philosopher had let loose "all his anger" on the physicist.[32] He was emboldened to write to Einstein directly. Einstein responded, exposing the philosopher's error. The recipient decided to publish it: "In short, Bergson forgets that spacetime simultaneity has an absolute character according to the Theory of Relativity."[33] By insisting on the "absolute" nature of relativity effects, Einstein tried to convince his readers that it was not necessary to think about relativity in terms of the *difference* between the two travelers. Einstein had not forgotten about Bergson.

4.4 The "Mistake" According to Bergson

One claim in Bergson's book is frequently cited as mistaken: that time is not altered according to the velocity of a system. He categorically stated that if a clock traveling close to the speed of light is later compared to a stationary clock, it "does not present a delay when it finds the real [stationary] clock, upon its return."[34] This claim, taken at first blush and in isolation, was completely at odds with the account of *time dilation* given by Einstein. In the foreword to his second edition, Bergson explained that the book's central message was to "explicitly prove that there is no difference, in what concerns Time, between a system in motion and a system in uniform translation."[35] "This sentence," noted the Bergson scholar and philosopher Élie During, "if taken literally, is evidently inacceptable for a physicist."[36] Yet Bergson's capitalization of Time signaled to readers that he was confronting a different concept from that of the physicists.[37]

Bergson, however, explained—on numerous occasions—that he did not understand by Time—and that is why he capitalized the term in the foreword—exactly the same thing as relativity physicists. In cases where one dealt exclusively with

---

[32] Miguel Masriera Rubio, "La verdad sobre Einstein," *La Vanguardia* (15 January 1925).

[33] Einstein to Masriera Rubio: "Kurz: Bergson vergass, dass raumzeitlische Koinzidenz auch nach der Relativitätstheorie absoluten Character hat." p. 1 of 2. Einstein to Masriera Rubio, 7 October 1925, Berlin. Scientific Correspondence, Folders M-Misc,1, Box 6, Einstein Archives.

[34] Appendix III in expanded second 1923 edition. Bergson, *Durée et simultanéité: á propos de la théorie d'Einstein*: 208.

[35] Foreword to the second expanded 1923 edition. Ibid., ix.

[36] Élie During, "Dossier critique: I. Notes," in *Durée et simultanéité: á propos de la théorie d'Einstein*, ed. Élie During (Paris: Quadrige/Presses Universitaires de France, 2009a), 253.

[37] For Bergson's strategic use of capital letters see Élie During, "Introduction au dossier critique," in *Durée et simultanéité: á propos de la théorie d'Einstein*, ed. Élie During (Paris: Quadrige/Presses Universitaires de France, 2009b), 237.

their notion of time, as in the case when only clocks were considered, or when dealing exclusively with physics and mathematics, or in the case when acceleration was involved, he fully accepted the conclusions of relativity scientists: "The truth is that the group of transformations discovered by Lorentz assures, in a general manner, the invariance of the equations of electromagnetism."[38]

Bergson allied his point of view on relativity with Henri Poincaré and Hendrik Lorentz, who never accepted essential aspects of Einstein's interpretation of relativity, and who, alongside Bergson, would be remembered for not having understood the theory or not being able to "let go" of old conceptions. "Poincaré never understood the basis of special relativity," explained Abraham Pais, Einstein's colleague and biographer.[39] Since then, others have echoed this familiar line. Walter Isaacson contends that "for his part, Poincaré seems never to have fully understood Einstein's breakthrough."[40] Dennis Overbye notes that "when he [Poincaré] finally addressed the issue of relativity squarely, in 1912, it was clear that he didn't understand it."[41]

Evidence shows that Poincaré accepted some of the most revolutionary implications of relativity, which he nonetheless attributed to Lorentz—not to Einstein. In a report on Lorentz's work, he repeated some of the same claims he had made earlier, when he had nominated Lorentz for the Nobel Prize. Poincaré explained how in the case of traveling, differing clocks, Lorentz had shown that it was impossible to claim one as correct and the other one as delayed. He explained how it was "impossible to detect anything other than relative velocities of bodies with regard to one another, and we should also renounce the knowledge of their relative velocities with regard to the ether as much as their absolute velocities." He concluded clearly: "This principle must be regarded as rigorous and not only as approximate."[42] That same year (1910) in a lecture in Gottingen, he framed the choice between Einstein's and Lorentz's interpretation of the theory as mainly a matter of taste. In his last significant statement on relativity, Poincaré did not even mention Einstein.[43] Instead, he chose to open his talk with an explanation of Bergson's philosophy.

Lorentz was in constant communication with Bergson. One day, when walking along the Seine, he brought up certain "objections" to Bergson's work. Bergson considered them carefully, and explained to Lorentz that his claim about the "two clocks" should not be taken literally. "It is just a manner of explaining oneself," so

---

[38] Bergson, *Durée et simultanéité: á propos de la théorie d'Einstein*: 24 n. 21.

[39] Pais, "*Subtle is the Lord...*"*: The Science and the Life of Albert Einstein*. 21.

[40] Walter Isaacson, *Einstein: His Life and Universe* (New York Simon and Schuster, 2007). 135.

[41] Dennis Overbye, *Einstein in Love* (New York: Penguin, 2000). 145.

[42] Henri Poincaré, "Rapport sur les travaux de H.A. Lorentz, ca. 31 January 1910," in *La Correspondance entre Henri Poincaré et les physiciens, chimistes et ingénieurs* (Basel: Birkhäuser, 2007), 438.

[43] Throughout, I use the term "relativity theory" and "theory of relativity" broadly to include the contributions of Lorentz and Poincaré. When referring to Einstein's particular contribution and to highlight his work from that of others, I use the phrase "Einstein's theory."

that he could get to "the depth of the matter."[44] Bergson explained why his philosophy was being received with such animosity on the part of physicists. He speculated that Einstein, along with numerous other physicists, simply did not understand him. To Lorentz, he offered a very negative view of Einstein:

> In general, relativity physicists have misunderstood me. They, by the way, frequently do not know my views except than through hearsay, by inexact and even completely false accounts. This is perhaps the case of Einstein himself, if what they say about him is true.[45]

Bergson later confided to another friend that Einstein could not comprehend him because "he is not that familiar with philosophy and especially with the French language."[46] He concluded that Einstein had probably not even "read my book" relying on second-hand accounts of "this or that French physicist who did not understand me, and who, not having the philosophical background needed to understand me, would remain impervious to my explanations."[47]

In the end, Bergson simply gave up trying to convince Einstein or his defenders. Their mutual misunderstanding was simply insurmountable. Referring to the accounts by Metz, he simply concluded: "The meaning of my thoughts, as that of my book, has completely escaped him. There is nothing I can do."[48]

4.5 Between Physics and Philosophy

What was Bergson's main point? Bergson had confronted the problem of time from many angles, most famously in *Matter and Memory* (1889) and later in *Creative Evolution* (1907). He agreed that clocks helped note simultaneities, but he did not think that our understanding of time could be based on them. He had already thought about this option, back in 1889, and had quickly discounted it: "When our eyes follow on the face of a clock, the movement of the needle that corresponds to the oscillations of the pendulum, I do not measure duration, as one would think; I simply count simultaneities, which is quite different."[49] In *Duration of Simultaneity* he came to the problem of time from a yet more sophisticated angle. He fully accepted the usefulness of clock time, but wanted to ask additional questions about it. He insisted that time was important for humans because of how it related to events that

---

[44] Bergson to Lorentz, 9 November 1924, Paris, in Henri Bergson, *Correspondances* (Paris: Presses Universitaires de France, 2002). 1119–1122 on p. 1122. Republished and annotated in Henri Bergson, *Écrits philosophiques*, ed. Arnaud Bouaniche, et al., Quadrige (Paris: Quadrige/Presses Universitaires de France, 2011). 556–559.

[45] Bergson to Lorentz, 9 November 1924, Paris, in Bergson, *Correspondances*: 1119–1122 on p. 1122. Republished and annotated in Bergson, *Écrits philosophiques*: 556–559.

[46] Jacques Chevalier, *Entretiens avec Bergson* (Paris: Plon, 1959). 69.

[47] Ibid.

[48] Henri Bergson, "Bergson à E. Peillaube," *Revue de philosophie* 24(July 1924a): 440.

[49] Henri Bergson, Essai sur les données immédiates de la conscience, ed. Arnaud Bouaniche, 9 ed. (Quadrige/Presses Universitaires de France, 2011). 80.

were relevant for them. One could not define time exclusively by reference to clocks, because clocks were made by humans *to time events that mattered to them and thus attracted their attention.*

Bergson argued that Einstein's theory of relativity was undergirded by a more basic sense of simultaneity, one which was not based on comparing events against clocks, but which would explain why clocks were invented and used in the first place. If this, much more basic, conception of simultaneity did not exist, then "clocks would not serve any purpose." "Nobody would fabricate them, or at least nobody would buy them," he argued. Yes, clocks were bought "to know what time it is," admitted Bergson. But "knowing what time it is" presupposed that the correspondence between the clock and an "event that is happening" was sufficiently *meaningful* for the person involved leading them to pay *attention* to it. That certain correspondences between events could be significant for us, while most others were not, explained our basic sense of simultaneity and the widespread use of clocks. Clocks, by themselves, could not explain either simultaneity or time, he argued.

If a sense of time more basic than that revealed by matching an event against a clock-hand did not exist, clocks would serve no purpose: "They would be bits of machinery with which we would amuse ourselves by comparing them with one another." Something different, something novel, something important, something outside of the clock itself needed to be included in our understanding of time. Only *that* could explain why we attributed to clocks such power: why we bought them, why we used them, and why we invented them in the first place. As the debate between the physicist and the philosopher unraveled in decades to come, the disagreement between their different conceptions of time appeared so inescapable that it was even traced back to the ancients, where Einstein's view was associated with Parmenides', and Bergson's with Heraclitus'.

Einstein tried to neutralize the philosopher by claiming that he did not deal with real things. According to Einstein, philosophy had been used to explain the relation between psychology and physics. "The time of the philosopher, I believe, is a psychological and physical time at the same time," he explained in Paris.[50] But relativity theory, by focusing on very fast phenomena, had shown just how off-the-mark psychological perceptions of time really were. Psychological conceptions of time, Einstein insisted, were not only simply in error, they just did not have an independent existence in reality. "These are nothing more than mental constructs, logical entities."[51] Because of the enormous speed of light, humans had "instinctively" generalized their conception of simultaneity and mistakenly applied it to the rest of the universe. Einstein's theory *corrected* this mistaken generalization. Instead of believing in an overlapping area between psychological and physical conceptions of time (where both were important although one was admittedly less accurate than the other), he argued that they were really two distinct concepts: a mental assessment (the psychological one) that was inadequate when compared to the only one "objective" concept: physical time.

---

[50] "La Théorie de la relativité: séance du 6 avril 1922," 363.

[51] Ibid., 364.

If Einstein could show that psychological assessments of time were essentially mistaken "mental constructs," then the task for philosophers was greatly reduced if not completely null. Bergson did not want to accept the role for philosophy that Einstein was giving to him. He was clear that his comments were not about physics, but that did not mean that they were about things that "did not exist" as Einstein seemed to imply in his phrase "*il n'y a donc pas un temps des philosophes.*"[52]

Bergson also did not accept that their disagreement was a technical matter in physics. None of *his* claims were meant to bear on physics: "The theory was studied with the aim of responding to a question posed by a philosopher, and no longer by a physicist." "Physics," he added, "was not responsible for answering that question."[53]

4.6 On the Relation Between the Special and General Theory

While the Michelson-Morley experiment played an important role in the *special* theory of relativity, three other experimental results were essential for the *general* theory (an explanation for the perihelion of Mercury, the bending of light rays by the sun, and the red-shift effect).[54] In the eyes of Einstein's supporters, these results proved the numerous virtues of Einstein's general theory. Were they (the "three classical tests" as they came to be known) not enough to blow Bergson's objections out of the water? In the decades that followed, the general theory received increasing support. A group of scientists working from new American large-scale observatories became "Einstein's jury" ruling in favor of the physicist.[55] A new way of understanding time in physics and astronomy gained prominence for the rest of the century.

Perhaps Bergson was mistaken because he focused only on the special theory? Bergson knew fully well that this approach had its limitations, but in his appendices and later publications he confronted claims that his book was irrelevant in the face of the success of the general theory. "In the special theory there is something that

[52] Ibid.

[53] Henri Bergson, "Les Temps fictifs et le temps réel," in *Durée et simultanéité: á propos de la théorie d'Einstein*, ed. Élie During (Paris: Quadrige/Presses Universitaires de France, 2009a), 425–426.

[54] Along with the Michelson-Morley experiment and the red-shift effect, historians and scientists alike often claim that the 1919 expedition was "one of the three 'classic' relativistic effects predicted by Einstein," Matthew Stanley, "An Expedition to Heal the Wounds of War: 1919 Eclipse and Eddington as Quaker Adventurer," *Isis* 94 (2003): 70. The red-shift effect was described by Einstein as a "crucial test" of the theory in 1916. To this day the bending of light rays by the sun is seen as a paradigmatic example of a crucial experiment. Albert Einstein, "The Foundation of the General Theory of Relativity," *CPAE* 6: 198. See entry "crucial experiment" in Nicholas Bunnin and Jiyuan Yu, *The Blackwell Dictionary of Western Philosophy* (Malden, MA: Blackwell, 2004).

[55] Jeffrey Crelinsten, *Einstein's Jury: The Race to Test Relativity* (Princeton: Princeton University Press, 2006).

demands the general theory," so by getting at the first theory he would touch on the second.[56] When readers were asked to consider clocks not only traveling away from each other, but meeting once again, how should they understand time? Should they use the algebraic equations of the special theory or the differential equations of the general one?[57]

If acceleration was taken into consideration, then the most astounding ways of describing Einstein's work broke down. To Bergson, the inclusion of acceleration proved that the two times were not equal in every sense: "So, if one wants to deal with real Times then acceleration should not create a dissymmetry, and if one wants for the acceleration of one of these two systems to effectively create a dissymmetry between them, then we are no longer dealing with real Times."[58] Acceleration was an inescapable mark of a difference in the clocks' travel itineraries. Since a difference existed, one that accompanied a difference in times, then their times were not equal in *every* sense.[59] After all, one would have the extraordinary experience of having done something different, being propelled to outer space and jolted back to return to Earth, while the other one comfortably remained at home. These differences were extraordinary, he argued, and physicists had no right to brush them aside and consider both observers as dealing with the same precious and contested entity, time.

If the dissymmetry due to acceleration was ignored, then Bergson was ready to concede to Einstein: "one could naturally say that [clocks traveling at different speeds] cannot run in synchronicity." In these cases "in effect Time slows down when speed increases." But for Bergson the introduction of acceleration proved that the times described by Einstein were not all equally real. "But what is this Time that slows down? What are these clocks that are not in synchronicity?"[60] These clocks were not equal in every way because one had gone through something that the other had not.

---

[56] Bergson, *Durée et simultanéité: á propos de la théorie d'Einstein*: 33.

[57] For a clear exposition of the twin paradox using only the special theory of relativity see Tim Maudlin, *Philosophy of Physics: Space and Time*, Princeton Foundations of Contemporary Philosophy (Princeton: Princeton University Press, 2012).

[58] Henri Bergson, "Les Temps fictifs et le temps réel," *Revue de philosophie* 24, no. 3 (1924b). Cited in Henri Bergson, "Les Temps fictifs et le temps réel," in *Mélanges* (Paris: Presses Universitaires de France, 1972 (1924)), 1443–1444.

[59] On the question of the symmetry/asymmetry of the relativity clocks see Jimena Canales, "Of Twins and Time: Scientists, Intellectual Cooperation, and the League of Nations," in *Neutrality in Twentieth-Century Europe: Intersections of Science, Culture, and Politics after the First World War*, ed. Rebecka Lettevall, Geert Somsen, and Sven Widmalm (New York: Routledge, 2012).

[60] "Appendix III" in Bergson, *Durée et simultanéité: á propos de la théorie d'Einstein*: 210.

4.7 Talking Past Each Other

How did Bergson deal with experimental results? In the opening sentence in *Duration and Simultaneity*, Bergson was very careful not to go against any facts of observation: "we take the formulas … term by term, and we find out to which concrete reality, to what thing perceived or perceptible, each term corresponds."[61] Bergson, in fact, wanted *more* not *less* weight placed on experiment and mathematics. He wanted to return to the results of the Michelson-Morley experiment—an experiment which was central in discussions of relativity theory.[62]

When describing the Michelson-Morley experiment, he again considered the connection between two clocks in terms of the exchange of electromagnetic signals. "How do we synchronize two clocks located at different places?" Two operators in charge of setting the clocks "communicate" with each other about the time by means of "optical signals, or more generally electromagnetic ones," "where a person in O sends a person in A a ray of light destined to be returned back." This procedure was equivalent to that of the Michelson-Morley experiment "with the difference, however, that mirrors have been replaced by people."[63] Both of these scenarios (one based on two individuals communicating with each other at a distance and the other one on the Michelson-Morley's experiment), according to Bergson, did not lead to Einstein's conclusions.

Bergson carefully considered the connection between the stationary observer and the traveling one in terms of electromagnetic communications. He imagined what the dialog between "Peter" and "Paul" could be as they separated. Bergson wrote it down, as if it were a script for a play. Peter, in Bergson's account of the story, says to Paul: "The moment you separated from me … your time swelled, your clocks disagreed." It was "obvious," according to Bergson what "Paul would reply." That everything was normal for him (Paul) and that it was Peter's system that had gone awry.

The conversation between the two observers, as narrated by Bergson, led nowhere. It consisted in back-and-forth repetition between the travelers. It was characterized by misunderstanding and mistrust. The conversation between Einstein and Bergson led to similar results.

Bergson considered a case where the travelers could be seen by "a supreme consciousness" capable of "communicating telepathically with both." That "consciousness" would indeed see the effects of time dilation. But "from the point of view of physics, that argument does not count."[64]

Throughout the rest of his book, Bergson explained how the laws of electromagnetism did not necessarily lead directly to Einstein's conclusions. When scientists turned to them as evidence for the theory of relativity, they assumed a concept of

---

[61] "Préface" in ibid., vi.

[62] Ibid., 1.

[63] Ibid., 9–10.

[64] Ibid., 91, 108.

communication so narrow that he could not possibly imagine how it would unambiguously compel scientists, or the famous travelers in the twin paradox, to agree with each other. Paul and Peter would go on disagreeing forever, never agreeing on the "paradoxes" of relativity theory. To fully investigate the topics of his interest—of how science passed from the abstract to the concrete, and from the conventional and the symbolic to the real—Bergson stressed aspects of the twin paradox that could *not* be explained simply be recourse to new forms of electromagnetic transmission and communication. He made it patently evident that Einstein, in his theory, made certain assumptions about the nature of communication.

4.8 Conclusion

Scientists' rhetorical, argumentative and persuasive practices are often held up as ideal examples of "reasonable discourse." The work of the philosopher Jürgen Habermas is often invoked to show how the road of "objective scientific truth" emerges from an " 'ideal speech situation,' the counterfactual ideal of a domination-free communication community."[65] In this view, bad science emerges when norms guaranteeing reasonable discourse are broken. The "freak of a [German] natural physics" and "Soviet Marxist genetics," explained Habermas, arose in cases of overt interference with the processes of consensual scientific deliberation.[66] The philosopher Helen Longino has drawn out a list to guard against the manipulation of science: "(1) there must be recognized avenues for the criticism of evidence, of methods, and of assumptions and reasoning; (2) there must exist shared standards that critics can invoke; (3) the community as a whole must be responsive to such criticism; (4) intellectual authority must be shared equally among qualified practitioners"[67] These parameters can give us hope that scientific controversies can be settled solely "on the basis of evidence." Philip Kitcher, author of *Science, Truth, and Democracy*, maintains that "neither the fact that major scientific controversies are protracted nor our inability to delineate a precise account of scientific evidence should undermine our confidence that the resolution of scientific debate on the basis of evidence is possible."[68]

Does the "ideal speech situation" described by Habermas actually lead to scientific consensus and to good science? Not everyone is an optimist. Being a scientist, according to the philosopher Mary Hesse, involves "the decision to enter a certain

[65] Gordon R. Mitchell, "Did Habermas Cede Nature to the Positivists?," *Philosophy and Rhetoric* 36, no. 1 (2003): 7.

[66] Jürgen Habermas, *Knowledge and Human Interests* (Cambridge: Polity, 1987). 315.

[67] Helen E. Longino, *Science as Social Knowledge: Values and Objectivity in Scientific Inquiry* (Princeton, N.J.: Princeton University Press, 1990). 76.

[68] Philip Kitcher, *Science, Truth, and Democracy*, Oxford Studies in Philosophy of Science (Oxford: Oxford University Press, 2001). 41.

form of life, that is, the scientific community of rational discourse."[69] But this ideal is rarely fulfilled. Hesse noted that "the conception of the ideal speech situation is certainly very strongly counterfactual. What then is its status? It is not something that is empirically realized in history, and perhaps is never realizable."[70] Habermas, agreeing with criticisms of this nature, nonetheless decided to hold on to this "fiction" for reasons that appear to be sentimental: "on this unavoidable fiction rests the humanity of intercourse among men who are still men."[71] The sociologist Milena Wazeck, in her study of Einstein's opponents, found that communication did not lead to scientific consensus. "The possibility of overcoming disagreement by communication and learning," she explains, in nothing other than an "optimistic view."[72]

Since the time of Galileo, science is replete with "rhetorical strategies of nondialog" as effective means for silencing opponents.[73] Working as a scientist involves working within a community of scientists that listen to "persuasive arguments" in favor or against of certain theories. But what happens when "unreasonable men" refuse to be or cannot be swayed by to the evidence that is presented to them? "The man who continues to resist after his whole profession has been converted has *ipso facto* ceased to be a scientist," explained Thomas S. Kuhn.[74]

When scientists communicate, they cannot help but communicate about communicating. What happens if we complement our studies of communication in science and our normative ideals of science with investigations into the actual non-communication strategies that actually *do* lead to consensus? This type of consensus comes at a steep price, excluding from our knowledge practices notions of communication connected to classical hermeneutics and exegesis that include investigations into meaning, understanding, intentionality, intertextuality, and affectivity and that have been relegated to a secondary status and circumscribed as part of the humanities. When talking about the time of the universe, it is about time to talk about who is doing the talking.

---

[69] Mary Hesse, "Habermas' Consensus Theory of Truth," *PSA: Proceedings of the Biennial Meeting of the Philosophy of Science Association* 1978 (1978): 382.

[70] Ibid., 381.

[71] Cited in T. McCarthy, "A Theory of Communicative Competence," *Philosophy of the Social Sciences* 3 (1973): 140; Hesse, "Habermas' Consensus Theory of Truth," 381.

[72] Milena Wazeck, "Marginalization Processes in Science: The Controversy about the Theory of Relativity in the 1920s," *Social Studies of Science* 43, no. 2 (2013): 185.

[73] Mario Biagioli, *Galileo, Courtier: The Practice of Science in the Culture of Absolutism* (Chicago: University of Chicago Press, 1993). 218. For communication breakdown in the case involving Herbert Dingle's objections to relativity, see Hasok Chang, "A Misunderstood Rebellion: The Twin-Paradox Controversy and Herbert Dingle's Vision of Science," *Studies in the History and Philosophy of Science* 24 (1994).

[74] Kuhn, *The Structure of Scientific Revolutions*: 159.

Selected Bibliography

1922. La Théorie de la relativité: séance du 6 avril 1922. *Bulletin de la Société française de philosophie* 22(3):349–370.

Arrhenius, Svante 1998. *Nobel lectures in physics (1901–1921)*. Singapore: World Scientific.

Barreau, Hervé. 1973. Bergson et Einstein: à propos de Durée et simultanéité. *Les Études bergsoniennes* 10: 73–134.

Benrubi, Isaac. 1942. *Souvenirs sur Henri Bergson*. Neuchâtel: Delachaux & Niestlé.

Bergson, Henri. 1924a. Bergson à E. Peillaube. *Revue de philosophie* 24: 440.

Bergson, Henri. 1924b. Les Temps fictifs et le temps réel. *Revue de philosophie* 24(3): 241–260.

Bergson, Henri. 1957. Écrits et paroles. In *Bibliothèque de philosophie contemporaine*, ed. Rose-Marie Mossé-Bastide. Paris: Presses Universitaires de France.

Bergson, Henri. 1972 (1924). Les Temps fictifs et le temps réel. In *Mélanges 1432–1449*. Paris: Presses Universitaires de France.

Bergson, Henri. 2002. *Correspondances,* ed. André Robinet. Paris: Presses Universitaires de France.

Bergson, Henri. 2009a. Les Temps fictifs et le temps réel. In *Durée et simultanéité: á propos de la théorie d'Einstein*, ed. Élie During, 417–430. Paris: Quadrige/Presses Universitaires de France.

Bergson, Henri. 2009b. *Durée et simultanéité: á propos de la théorie d'Einstein,* ed. Élie During and Frédéric Worms, 4th ed. Paris: Quadrige/Presses Universitaires de France.

Bergson, Henri. 2011 (1889). *Essai sur les données immédiates de la conscience,* ed. Arnaud Bouaniche and Frédéric Worms, 9th ed. Paris: Quadrige/Presses Universitaires de France.

Bergson, Henri. 2011. *Écrits philosophiques,* ed. Arnaud Bouaniche, Arnaud François, Élie During, Frédéric Fruteau de Laclos, Frédéric Keck, Stéphane Madelrieux, Camille Riquier, Ghislain Waterlot and Frédéric Worms. Quadrige. Edited by Frédéric Worms. Paris: Quadrige/ Presses Universitaires de France.

Biagioli, Mario. 1993. *Galileo, courtier: The practice of science in the culture of absolutism.* Chicago: University of Chicago Press.

Biezunski, Michel. 1991. *Einstein à Paris: le temps n'est plus....* Vincennes: Presses Universitaires de Vincennes.

Bunnin, Nicholas, and Jiyuan Yu. 2004. *The blackwell dictionary of western philosophy*. Malden: Blackwell.

Canales, Jimena. 2012. Of twins and time: Scientists, intellectual cooperation, and the league of nations. In *Neutrality in twentieth-century Europe: Intersections of science, culture, and politics after the First World War*, ed. Rebecka Lettevall, Geert Somsen, and Sven Widmalm, 243–270. New York: Routledge.

Canales, Jimena. 2015. The media of relativity: Einstein and communications technologies. *Technology and Culture* 56(3): 610–645.

Chang, Hasok. 1994. A misunderstood rebellion: The twin-paradox controversy and Herbert Dingle's vision of science. *Studies in the History and Philosophy of Science* 24: 741–790.

Chevalier, Jacques. 1959. *Entretiens avec Bergson*. Paris: Plon.

Clark, Ronald William. 1971. *Einstein: The life and times*. New York: World Publishing.

Crelinsten, Jeffrey. 2006. *Einstein's jury: The race to test relativity*. Princeton: Princeton University Press.

du Noüy, Pierre Lecomte. 1936. *Biological time*. London: Methuen and Co.

During, Élie. 2009a. Dossier critique: I. notes. In *Durée et simultanéité: á propos de la théorie d'Einstein*, ed. Élie During, 219–244. Paris: Quadrige/Presses Universitaires de France.

During, Élie. 2009b. Introduction au dossier critique. In *Durée et simultanéité: á propos de la théorie d'Einstein*, ed. Élie During, 219–244. Paris: Quadrige/Presses Universitaires de France.

Einstein, Albert. 1997 (1916). The foundation of the general theory of relativity. In The Berlin Years: Writings: 1914–1917, Collected Papers of Albert Einstein, 6: 147–200. Princeton: Princeton University Press.

Fölsing, Albrecht. 1997. *Albert Einstein: A biography*. New York: Viking.

Frank, Philipp. 1947. *Einstein, his life and times*. Trans. George Rosen. New York: A.A. Knopf.

Genovesi, Angelo. 1998. Henri Bergson: Lettere a Einstein. *Filosofia* 49(1): 3–39.

George, André. 1930. Bergson et Einstein. *Les documents de la vie intellectuelle*: 52–64.

Gunter, P.A.Y. (ed.). 1969. *Bergson and the evolution of physics*. Knoxville: The University of Tennessee Press.

Habermas, Jürgen. 1987. *Knowledge and human interests*. Cambridge: Polity.

Hafele, J.C., and R.E. Keating. 1972. Around-the-world atomic clocks: Predicted relativistic gains. *Science* 177: 166–170.

Hanna, Thomas (ed.). 1962a. *The Bergsonian heritage*. New York: Columbia University Press.

Hanna, Thomas. 1962b. Introduction. In *The Bergsonian heritage*, ed. Thomas Hanna, 1–31. New York: Columbia University Press.

Hermann, Armin. 1996. *Einstein: Der Weltweise und sein Jahrhundert; eine Biographie*. München: Piper.

Hesse, Mary. 1978. Habermas' consensus theory of truth. *PSA: Proceedings of the Biennial Meeting of the Philosophy of Science Association* 1978: 373–396.

Isaacson, Walter. 2007. *Einstein: His life and universe*. New York: Simon and Schuster.

Kitcher, Philip. 2001. *Science, truth, and democracy*, Oxford studies in philosophy of science. Oxford: Oxford University Press.

Kuhn, Thomas S. 1970. *The structure of scientific revolutions*, 2nd ed. Chicago: University of Chicago Press.

Lakatos, Imre. 1970. Falsification and the methodology of scientific research programmes. In *Criticism and the growth of knowledge*, Studies in logic and the foundations of mathematics, ed. Imre Lakatos and Alan Musgrave. Cambridge, UK: Cambridge University Press.

Langevin, Paul. 1911. L'evolution de l'espace et du temps. *Scientia* 10: 31–54.

Latour, Bruno. 2003. Do you believe in reality? News from the trenches of the science wars. In *Philosophy of technology: The technological condition*, ed. Robert C. Scharff and Val Dusek, 126–137. Malden: Blackwell Publishers.

Ledesma, Ramiro. 1929. Actualidad. Filosofía, Ciencia: Andre Metz y la Geometría Euclidiana. *La Gaceta Literaria* 54: 2.

Longino, Helen E. 1990. *Science as social knowledge: Values and objectivity in scientific inquiry*. Princeton: Princeton University Press.

Maritain, Jacques. 1926 (1924). *Réflexions sur l'intelligence et sur sa vie propre*. Paris: Nouvelle Libraire Nationale.

Markosian, Ned. 2014. Time. In *Stanford encyclopedia of philosophy*, ed. Edward N. Zalta. URL = <http://plato.stanford.edu/archives/spr2014/entries/time/>.

Masriera Rubio, Miguel. 15 January 1925. La verdad sobre Einstein. *La Vanguardia*: 9.

Maudlin, Tim. 2012. *Philosophy of physics: Space and time*, Princeton foundations of contemporary philosophy. Princeton: Princeton University Press.

McCarthy, T. 1973. A theory of communicative competence. *Philosophy of the Social Sciences* 3: 135–156.

Metz, André. 1923. *La Relativité: Exposé dans formules des théories d'Einstein et réfutation des erreurs contenues dans les ouvrages les plus notoires*. Paris: Etienne Chiron.

Metz, André. 1924a. Le Temps d'Einstein et la philosophie: à propos de l'ouvrage de M. Bergson, Durée et simultanéité. *Revue de philosophie* 31: 437–439.

Metz, André. 1924b. Replique a Bergson. *Revue de philosophie* 31: 437–439.

Metz, André. 1959. Bergson, Einstein et les relativistes. *Archives de Philosophie* 22: 369–384.

Mitchell, Gordon R. 2003. Did Habermas cede nature to the positivists? *Philosophy and Rhetoric* 36(1): 1–21.

Mossé-Bastide, Rose-Marie. 1955. *Bergson éducateur*. Paris: Presses Universitaires de France.

Neffe, Jürgen. 2007. *Einstein: A biography*, 1st American ed. New York: Farrar, Straus, and Giroux.

Overbye, Dennis. 2000. *Einstein in love*. New York: Penguin.

Pais, Abraham. 1982. *"Subtle is the Lord...": The science and the life of Albert Einstein*. Oxford: Clarendon.
Papanicolaou, Andrew C., and P.A.Y. Gunter. 1987. *Bergson and modern thought: Towards a unified acience. Models of scientific thought*. Chur: Harwood Academic Publishers.
Poincaré, Henri. 2007. Rapport sur les travaux de H.A. Lorentz, ca. 31 January 1910. In *La Correspondance entre Henri Poincaré et les physiciens, chimistes et ingénieurs*. Basel: Birkhäuser.
Prigogine, Ilya. 1971. Evolution of physics: Review of Bergson and the evolution of physics, edited and translated by of P.A.Y. Gunter. *Nature* 234: 159–160.
Sokal, Alan D., and J. Bricmont. 1997. *Impostures intellectuelles*. Paris: O. Jacob.
Sokal, Alan D., and J. Bricmont. 1998. *Fashionable nonsense: Postmodern intellectuals' abuse of science*. New York: Picador.
Stanley, Matthew. 2003. An expedition to heal the wounds of war: 1919 eclipse and eddington as quaker adventurer. *Isis* 94: 57–89.
Telkes-Klein, Eva. 2007. Meyerson dans les milieux intellectuels français dans les années 1920. *Archives de Philosophie* 70(3): 359–373.
Wazeck, Milena. 2013. Marginalization processes in science: The controversy about the theory of relativity in the 1920s. *Social Studies of Science* 43(2): 163–190.

Part II
Transience and Experience

Chapter 5
Some Cosmological Implications of Temporal Experience

Barry Dainton

Abstract The claim that we directly apprehend change and succession in our ordinary experience is phenomenologically plausible – after all, we certainly *seem* to, at least over short intervals. However, there are those who hold that any attempt to grant consciousness temporal breadth falls into incoherence. I argue here that this is wrong, and defend the "extensional" conception of temporal experience over its "retentional" rival. I then argue that this means that reality itself must be dynamic, simply because experience is dynamic, and experience is part of reality. The precise extent to which the dynamism we find in our experience impacts on the rest of the universe depends on the (much contested) relationship between consciousness and the rest of the universe; on some views it will be trivial, on others immense. Of the different metaphysical accounts of the nature of time currently on offer, Presentism is (arguably) the most dynamic. However, Presentism – at least in its standard guise – looks to be irreconcilable with the extensional account of temporal experience. The way forward, I suggest, is to adopt a modified form of Presentism. I conclude by examining the implications of this view of experience for the claim that we might very probably be short-lived "Boltzmann Brains".

Keywords Block universe • Presentism • Temporal experience • Motion perception • Specious present • Extensional model • Retentional model • Extensional presentism • Synchronic and diachronic co-consciousness • Transitivity • Boltzmann brains

---

B. Dainton (✉)
Department of Philosophy, University of Liverpool, Liverpool, UK
e-mail: bdainton@liverpool.ac.uk

Y. Dolev, M. Roubach (eds.), *Cosmological and Psychological Time*, Boston Studies in the Philosophy and History of Science 285,
DOI 10.1007/978-3-319-22590-6_5

5.1 Introduction

An article in a 2013 *New Scientist* entitled "The now delusion" by Michael Slezak provides provocative food for thought. Sub-titled "Do past, present and future exist only inside our heads?", the piece starts off in a familiar vein:

> Imagine standing outside the universe. Not just outside space, but outside time too. From this spectacular vantage point, you gaze down upon the universe. At one end you see its beginning: the big bang. At the other, you see … whatever it is that happens there. Somewhere in the middle is you, a miniscule worm: at one end a baby, the other end a corpse. From this impossible perspective, time does not flow, and there is no "now". Time is static. Immutable. Frozen.
>
> Fantastical as it seems, for most physicists today the universe is just like that. We might think of time flowing from a real past into a not-yet-real future, but our current theories of space and time teach us that past, present and future are all equally real – and fundamentally indistinguishable. Any sense that our "now" is somehow special, or that time flows past it is an illusion we create in our heads.

What Slezak is describing here is "the block universe", a universe which lacks a privileged present – let alone a privileged present which is steadily advancing towards the future, second by second – a universe where time (in this sense at least) does not pass. As for the "current theories of space and time" which teach us that we inhabit a passage-free universe where all times and events are equally real, the most influential by far is Einstein's special theory of relativity. It is difficult to hold that the present has any deep metaphysical significance if subjects in motion relative to one another will legitimately regard different sets of events *as* present. Since Einstein's theory relativizes the present (or distant simultaneity) in precisely this way, it is an extremely hostile environment for metaphysical accounts of time which hold that the present is a steadily advancing universe-wide interface between existence and non-existence. Needless to say, it is also hostile to anything resembling most people's pre-philosophical pre-conceptions regarding the nature of time.

The story so far is all too familiar, at least to philosophers with an interest in the nature of time, but the tale Slezak goes on to relate tell is less well known. "Physics" he tells us "has killed time as we know it. The question is: do we need it back?" In their long-running – but still unsuccessful – quest to unify quantum mechanics with relativity, some physicists are now arguing that the stumbling block in reconciling our best theory of micro-world with our best theory of the macro-world is the very block conception of time that has gone largely unquestioned – in physics circles – for the past century. As a consequence, Slezak argues, conceptions of time that have been taken seriously only by a handful of metaphysicians are now being given serious attention by prominent physicists. Since these conceptions of time incorporate both a privileged present and passage, they are much closer to our common sense conception of time than the frozen block universe.

One of these alternatives, favoured by physicist George Ellis, is the "growing" or "evolving" block universe. According to this view of time, the past is real, the future unreal, and time advances by virtue of a process of creation: new presents are continually coming into existence, moment by moment. What we call *the now* is simply the most recently created slice of reality. Ellis champions the growing block model because he thinks that physics itself requires us to recognize the existence of a privileged present. We know from physics – or so Ellis argues – that future events are inherently unpredictable. We can assign probabilities to this or that event happening at some future time *t*, but we will only know for certain what happens at *t* when *t* becomes present. Hence "Things could have been different, but second by second, one specific evolutionary history out of all the possibilities is chosen, takes place, and gets cast in stone" (Ellis 2006: 1812–3). Although this indeterminacy applies, or so Ellis argues, to the behaviour macroscopic objects, such as wine glasses that are about to shatter, or living organisms, it also – and most profoundly – exists at the micro-level, where quantum mechanics rules. From the quantum mechanical perspective, the present is the "transition region in which quantum uncertainty changes to classical definiteness" (Ellis and Rothman 2010: 989).

Of course, the interpretation of quantum mechanics is a controversial issue, to put it mildly. Those who deny that an object's wave function ever collapses – as proponents of "many worlds" interpretations do – will find no room for the kind of present Ellis favours. It should also be noted that Ellis agrees with the mainstream block theorists in holding that Einstein's special relativity rules out a single universe-wide present. Accordingly, what evolves is not the universe as a whole, but the worldlines of individual particles:

> spacetime is extending to the future as events develop along each world line in a way determined by the complex of causal interactions … There is no unique way to say how this happens relatively for different observers …. to describe it overall, it will be convenient to choose specific time surfaces for the analysis but these are a choice of convenience rather than necessity (2006, sec. 4.5)

With his evolving block universe Ellis considerably narrows the gap between the physicist's conception of time and the common sense conception, but a gap still remains.[1]

There are other physicists who would further narrow the gap. Like Ellis, Lee Smolin has also come round to the view that progress in physics requires the abandonment of the block conception of time. But rather than settle for the half-way house that is the growing block view, in his recent Time Reborn: From the Crisis in Physics to the Future of the Universe (2013a), Smolin argues that the block should be dispensed with completely. In its stead, Smolin advocates a version of *Presentism*, according to which both the past and the future are wholly unreal. The past and future may both be unreal, but there are nonetheless differences between them: there are no facts about what the future may or may not contain, but there are facts about the past. Time passes on this view too: it consists of one momentary (or very brief) slice of reality – one present – giving way to another, then another, then another.

[1] The gap is in fact larger than this. For the sake of brevity I have not mentioned an additional complication: "[when] we contemplate a crystallizing nature for the emergence of spacetime: not all features become fixed at the same, and post-selection of previous events is possible. Potentiality changes to actuality at each quantum measurement process, but some potentialities may remain undecided even as others have transmuted to definiteness." (Ellis and Rothman 2010: 18)

Crucially, the present is not confined to the world-lines of individual particles, as on Ellis' account. Smolin's present is both objective and universe-wide.

Hasn't Einstein shown that an all-encompassing present of this sort has no place in physics? Smolin thinks not. So far as he is concerned, the key development is a new reformulation of general relativity known as "shape dynamics", developed by Julian Barbour and his colleagues. Previous attempts to incorporate absolute simultaneity into general relativity have proved empirically inadequate, but this is no longer the case. Shape dynamics makes precisely the same empirical predictions as standard general relativity while preserving a single universal present. It does so by relativizing *size*, rather than time. As a consequence of this, observers moving relative to one another will all agree on *when* two events occur, they will disagree on how large the relevant objects are – and there is no fact of the matter as two who is correct.

As Slezak also relates in his article, these recent developments are also – not surprisingly – proving controversial. Philosophers (such as Huw Price) who believe that dynamic conceptions of time are metaphysically incoherent are unmoved. As are physicists – such as Sean Carroll, for example – who believe that the block universe is not in fact past its sell-by date. Nonetheless, it is a striking development that scientists working at the forefront of physics and cosmology are starting to take a serious interest in the different metaphysical conceptions of time that philosophers have been debating for centuries (if not longer). At the very least, it means that philosophers who combine an interest in temporal metaphysics with a respect for contemporary physics and cosmology have more room for manoeuvre. As we wait for further scientific developments, we no longer need to work on the assumption that the block universe is the only conception of time with respectable scientific credentials.

In what follows I will not be attempting to assess the scientific considerations which have led physicists such as Ellis and Smolin to embrace alternatives to the block view. My aim is more modest. What I want to focus on are the implications of our temporal *experience* for our metaphysical theorizing about time. In one respect, these are less than momentous, for I do not think that the temporal features of our experience decisively favour one particular metaphysical conception of time over another. However, in other respects they are by no means entirely negligible, or so I will be arguing.

5.2 The Multiplicity of Motion

Zeno's paradox of the arrow is well known. We start by assuming that time and space are continuous, and composed of an infinitely dense succession of durationless instants and extensionless points respectively. As it moves along its flight-path, the tip of the eponymous (and very sharp) arrow passes through every point on this path, occupying each for just an instant. At each of these instants, the arrow as a whole occupies a volume of space that is precisely equivalent to its own volume.

Evidently, at each of these instants the arrow is at complete rest, for it does not change its location in space – there is simply no time for it to do so. Since the arrow is not in motion at any time (or instant) during its entire flight, it is *never* in motion. The arrow, it seems, cannot fly, and yet it does. Since the same applies to all moving objects, it seems that motion cannot occur … yet it surely does.

The standard contemporary response to this conundrum is also reasonably familiar. In his *Principles of Philosophy* (1903: 308) Russell argued that Zeno's argument for the impossibility of motion fails, but in an instructive way. We naturally think that moving objects differ in some intrinsic way from objects which are stationary. What Zeno's argument shows is that this is not in fact the case. At any given instant, there is no difference at all between an arrow that is moving at 100 kmh, and one that is moving at 0 kmh. Like any other object, a moving object exists at a succession of durationless instants. It is not moving *during* these instants – it is entirely motion-free, static. However, it is properly regarded as being in motion if it is at different spatial locations at neighbouring times. An object which doesn't change its location at instants surrounding a given time *t* is not at motion at *t*; if the object in question varies its location, then it is in motion, even though its intrinsic condition or state at *t* is indistinguishable from that of its motionless counterpart. Russell summarized this "at-at" account of motion thus: "Motion consists merely in the fact that bodies are sometimes in one place and sometimes in another, and that they are at intermediate places at intermediate times" (1917: 84).

In one sense the at-at account gives us everything we might want of an account of the nature of motion: after all, no one would deny that an object which is continuously occupying different spatial locations during a given interval is moving, whereas an object which remains stuck at the same place is not. However, it remains the case that the claim that objects in motion and objects at rest do not differ one iota in their intrinsic properties does not ring true; intuitively, we feel there *is* a difference, a deep and radical one. As for *why* we feel like this, there is probably no single or straightforward answer, but a part – a substantial part, I suspect – of the answer is not hard to find: motion is a property of objects that we can directly perceive, and moving objects simply *look* very different from objects that are motionless.

Some objects that are in motion – a helicopter's propeller, or a bullet speeding on its way to its target – are moving too quickly for us to see them. Other objects are in motion, but moving too slowly; the stars in the night sky, the growth of a fingernail, or the tectonic movement of a continent, all fall into this category. Some objects are both too small and moving too fast: atoms and molecules, for example. But when it comes to the world of ordinary medium-sized things, motion is not only all around us, it can manifest in manifestly different ways.

More or less dramatic examples of the distinctive animated character of directly perceived motion are not hard to find. To illustrate the vast difference between noticing that something *has* moved (or otherwise changed) from seeing something actually *moving* or *changing* C.D. Broad (1923: 351) invited us to consider what it's like to observe the second-hand of a watch or a flickering flame. He might have as easily invited us to compare the experience of looking at a still photograph of a waterfall with looking at a waterfall itself (or a video recording of one). The turbulent plunging of the water – it's continuous turning twisting flow – are as vividly present in our experience as its colour or shape. Or consider what it's like to see a bird fly across the sky, or feel water running down one's back, or see the world viewed from a train window sliding by – or watch an arrow fall from the sky.

Of course, quite how one perceives a moving arrow will depend on how fast it is moving – and likewise the water in a waterfall. If the arrow has been propelled at very high speed by a high-powered bow, it might appear as little more than a blur. But if it is travelling less quickly (or is travelling quickly but viewed via slow-motion video) it will be seen clearly and cleanly without a hint of blur, but very definitely moving nonetheless. The difference between experiencing *slow + clear* motion and *faster + blurry* moving objects can easily be illustrated: just try moving a pen back and forth in front of your eyes at different speeds. When moving slowly – e.g. each movement from right to left or left to right takes a couple of seconds – the pen is seen *as moving*, but also as clearly and cleanly delineated. Double or triple the speed, and the pen can still be seen moving back and forth but it appears as a streak, rather than a cleanly demarcated object; move it faster still and you will see it dissolve into nothing but *pure blur*.

Appreciating the extent to which motion features so prominently and distinctively in our everyday experience helps explain why the at-at account can seem less than the full story. According to the latter, a moving object is simply an object which exists at a sequence of different spatial locations in an entirely static and *motionless* fashion. Motion as-we-perceive-it is utterly different. When we look (say) at a waterfall, we see nothing but turbulence, turmoil and flow: there is typically *nothing* at all in our experience of the water that static and immobile. Quite generally, experienced motion possesses an inherent dynamism that is hard to reconcile with the sequence of immobilities offered by the at-at theory.[2]

To register the difference between these different forms (or aspects) of movement, let us refer to the inherently dynamic motion-as-it-features-in-our-experience as phenomenal or *P-motion*, and the non-dynamic motion as characterize by the at-at theory as non-phenomenal, or *NP-motion*. The drawing of this distinction sheds some useful light on the at-at theory. If this is intended to be a full and complete account of motion in our universe in all its forms, then given that P-motion exists and is not captured by the at-at view, the latter is clearly inadequate. However, if the at-at theory is simply supposed to provide an account of motion as it exists in material bodies, and *not* our experience, then its prospects look brighter. If material bodies are entirely lacking in phenomenal properties, as many suppose, then they will not possess the dynamic intrinsic qualities associated with P-motion. In this less ambitious guise, the at-at view may be correct; at the very least it is not obviously false.

---

[2] A point forcefully put by Bergson who argued that "with immobility set beside immobility, even endlessly, we could never make movement" (1911: 307). In fact, motion as we encounter it in perception is *indivisible*, i.e. not composed of a sequence of static states, or so Bergson held. Since Bergson subscribed to a direct realist view of perception, he held that the forms of motion we encounter in our perceptual experience are features of the external world, just as they seem to be. I will be returning to this theme in Sect. 5.5.

There is, however, a second point to note that is of some importance. Recall the terms with which Slezak characterized the block universe in the passage I quoted at the start: it is "static", "immutable", "frozen". As should by now be apparent, this is not the whole truth. Let's suppose our universe is of the block variety, and there is no such thing as a privileged present, or temporal passage as this is usually construed (e.g. in terms of a present advancing towards the future). It is nonetheless false to claim that our universe in its *entirety* is static and immutable. Why? Because irrespective of what else it contains, we know that it also contains P-motion – a form of motion which (as we also know) is intrinsically highly dynamic. Now, it may be – as many philosophers suppose – that our experience is in some respects misleading with regard to the properties it represents the world as possessing. Perhaps the material objects that we perceive in our everyday surroundings are devoid not only of phenomenal colour, but of all trace of phenomenal motion. Even so, it remains the case that P-motion exists as a *real feature of our experience*. And since our experience is as much a part of the concrete fabric of universe as anything else – or so I think it plausible to suppose, and will be assuming here – so too is P-motion and its intrinsic dynamism. Slezak himself suggests that "Any sense that our 'now' is somehow special, or that time flows past it is an illusion we create in our heads". Perhaps so. But given that our experience is itself part of the universe, the dynamism that exists in our experience is *also* an integral part of our universe – even if it turns out to be confined *to* our experience.

Needless to say, when it comes to the issue of whether, and the extent to which, our perceptual experience misrepresents the world we perceive, there are very different views, and the same goes for the relationship between consciousness in all its forms and the rest of the universe. Not surprisingly, the different positions on these issues have significant implications for the extent to which the dynamic features we encounter in our experience are representative of the wider universe. I will be returning to this issue shortly. But first there are some preliminaries to be taken care of.

5.3 Temporal Experience

I have been suggesting that the phenomenological evidence is utterly unambiguous: motion is something we directly apprehend in our immediate experience, and when we apprehend motion, our experience has an intrinsically dynamic *sui generis* character. When we see an arrow flashing through a forest, our experience is of an object *in motion*; it does not consist of a succession of static momentary snapshots of a static, motion-free object.

So far so plausible, phenomenologically at least, but there remains a potential problem. Motion takes time. If I am directly aware of an object moving from A at t_1 to B at t_2, then my awareness must extend over this temporal interval. But how can this be? Isn't our direct experience confined to the present? Isn't the present momentary?

German psychologists working on various aspects of temporal experience in the ninenteenth century encountered this issue, and quickly discerned the outlines of a promising solution. Yes, the present strictly so-called is momentary, but the *psychological* or *experiential* present is not: it extends over a brief interval of time, sufficient to allow motion (and other forms of change or persistence) to be directly apprehended in consciousness. This interval goes by a variety of names – it is often labeled the "specious present" – but also, and more importantly, it can be defined in different ways. In his *Principles of Psychology*, William James construed the specious present in a broad way, intending it to encompass both the apparent span of immediate awareness, and also periods of time which are cognitively privileged in any of several ways. We have more accurate memories of events witnessed six seconds ago, for example, than we have of events witnessed six minutes or six hours ago; we are able to estimate the duration of shorter intervals more accurately than their longer counterparts. If, however, we construe the period in question more narrowly, and take the experiential present solely to refer to periods that are can seemingly be *experienced* as temporally extended wholes, then we are dealing with only quite brief periods, of probably less than a second. (To illustrate: if you click you fingers once, then again a second or so later, you aren't still *hearing* the first click when the second is heard; however, the first click – itself a temporally extended auditory item – is experienced as a whole.) It is only with this narrower construal of the experiential present that we shall be concerned with henceforth.

Although doctrine of the specious present meets a real need, it has also been provoked suspicion, and some have doubted whether the notion is really intelligible. The claim that the doctrine is fundamentally incoherent has been recently voiced by Dolev. One of his complaints runs thus:[3]

---

[3] In the same article Dolev also argues that if we adopt (as he recommends) a contextualist conception of the present, the problem of understanding how motion can be perceived simply evaporates, and the specious present in all its forms is superfluous to requirements. On this view, depending on our current concerns, the present can contract to an instant or expand to a century (or more); since some of the events we perceive involve objects that are in motion, *of course* we can perceive this motion, provided we expand the present to encompass a relevant temporal interval.

> The experience of motion does not consist of a perception of succession nor is it derived from a succession of perceptions. Succession plays no role in the perception of motion, or for that matter, in motion itself. Seeing the cat cross the lawn does not consist of having a succession of perceptions, nor of perceiving succession. The experience simply does not break down to components that have to then be reunited. Hard as we try, when we scrutinize our experience we seek in vain for past bits that somehow coalesce with present ones to form the perception of motion. (2014: 41)

Dolev certainly gets our ordinary language use of "the present" right, but I don't believe it is remotely plausible to hold that the *experiential* present – the brief period we directly apprehend – varies with context (not, that is, unless Dolev's experience is very different from mine). Nor can I see how motion can be directly perceived unless our awareness *is* capable of extending (or its seeming to) over a brief interval. If I am directly aware of an object moving from A to B, then this movement must exist *as a temporally extended whole* in my experience. If this were not the case, I would not be perceiving the movement per se, but merely seeing the object at B (say), and inferring that it had moved. The different phases of a perceived movement must exist together in my consciousness, or I wouldn't be *perceiving* the movement at all.

> The perception of change and motion seem to confront us with an antinomy: it can be achieved only if temporally distant contents are apprehended together. The solution is to create a specious zone that is not subject to the conceptual and logical constraints our descriptions of the world itself must respect, and let perception of change take place there. (2014: 38)
>
> … the specious present is a contraption designed to perform magic, namely, conjugate what are unconjugable, distant events for example, blending together the distinct notes of the sequence A-B-C-D into a tune. The specious present was invented so that there is something that will tolerate what actual motion and change do not, the coexistence of phases of the succession. The notes A-B-C-D are temporally distant and so cannot co-exist in reality, but they can be co-conscious in the specious present … How does this magic happen? How does the temporal gap … vanish when they arrive at the theatre in which they are heard, the specious present? (2014: 37)

For Dolev, in effect, the specious present is *doubly* specious. It not only misleads us with regard to the reality of time by implying that the present has duration when in fact it is durationless, but it is also misleading with regard to the temporal *appearances*: proponents of the doctrine maintain that our immediate awareness is temporally extended, when in reality it is not.

What Dolev finds problematic is the way in which the specious present, if *per impossible* it were to exist, would permit the co-existence of things which (in reality) do not and cannot co-exist. The specious present is thus a fiction introduced to perform what is logically impossible, to unite what cannot be united. But is there really a problem here?

We can agree that it is incoherent to hold that "X and Y co-exist at the same time; it's also the case that X and Y do not co-exist at that time". It is similarly incoherent to maintain that "X and Y co-exist in the same universe, and X and Y do *not* co-exist in the same universe". But why think the specious present theorist is committed to either of these obviously contradictory claims? All these theorists are claiming is that successions which *do* in fact occur in reality are experienced as successions. What is impossible or problematic about that?

In fact, there are very different accounts of the specious present itself, and the differences between these are very much relevant to Dolev's objection. One prominent approach, which can be traced back to the work of Brentano in the latter half of the ninenteenth century, and subsequently Husserl, holds that specious presents lack any objective temporal extension. The phenomenal content of a specious present *seems* to possess temporal extension – it features a brief period of movement or change – but this dynamic content is encapsulated in an episode of experience that is momentary (or does not possess any significant temporal depth). We can call this the *Retentional* model, to register Husserl's influence.[4] This model can be – and has been – developed in different ways, but for present purposes many of these variations are irrelevant.

[4] Husserl took the vast bulk of a specious present to be made up of "retentions", which (roughly) are representations of the recent past.

The main alternative – which elsewhere I have called the *Extensional* model – can be traced by to William Stern (1897), a near contemporary of Husserl's.[5] On this view, a specious present is also a unified episode of experience whose contents appear to be dynamic, but rather than being momentary it extends a short distance through time, in much the way it seems to. As a temporally extended episode of experience, an Extensional specious present has briefer experiences as phases or parts; a Retentional specious present, in contrast, does *not* possess briefer experiences as parts. So although a Retentional specious present will seem to possess successive phases, these phases co-exist in an objectively momentary (or near) momentary conscious state. In sharp contrast, the successive phases of an Extensional specious present are *not* confined to a momentary conscious state: they are parts of a temporally extended episode of consciousness.

Needless to say, there is a good deal more to be said about both these models, but as will already be clear, Dolev's objection impacts upon them in very different ways. So far as the Retentional specious present is concerned, if this is construed in the usual way, it consists of phenomenal contents which exist simultaneously in objective time. Hence if a specious present of this type contains a succession of very brief tones C-D-E, although these contents seem to extend through time, in fact they coexist within a single momentary conscious state. Given this, the claim that the specious present requires contents that do not in fact coexist to coexist is simply false. The contents of this kind of specious present clearly *do* coexist. They coexist in the most straightforward way possible: in a single momentary state.[6]

The situation with regard to the Extensional model is very different. Here the contents C-D-E exist at different times, both subjectively and objectively. Since C exists in a single unified conscious state which includes both D and E, these experience (or experience-phases) must also coexist. It cannot be the case that C is experienced together with D if C does not exist – does not figure in reality *at all* – by the time D is experienced. If C is co-conscious with D, then both C and D must exist, despite the fact that they each occur at different times.

Now, Dolev insists that it is absurd to suppose that contents such as C and D – i.e. contents which do not co-exist simultaneously – can be experienced together. But why, exactly? There are only two obvious possibilities. One might think that phenomenal unity, of the kind found in unified conscious states such as (*ex hypothesi*) specious presents is by its very nature confined to states which lack temporal extension. In which case, a version of the Retentional account of temporal experience must be correct. There is a second, but not necessarily unrelated, reason why one might think that co-consciousness required simultaneous co-existence. This requirement would, presumably obtain if Presentism were true, i.e. if only the present moment exists, and the past and future are both wholly unreal. The confinement of

[5] For further discussions of, and elaborations on, this approach to temporal experience see Dainton 2006, 2008, 2010a, b, 2011b, 2014a.

[6] Although Husserl adopted the Retentional model, his position is more nuanced, given the framework of transcendental phenomenology that he came to adopt. The latter dispenses with the physical world as this is usually conceived.

phenomenal unity to an unextended present may be a consequence of the nature of time, rather than the nature of consciousness and phenomenal unity *per se*.

With respect to this first of these reasons, while it is certainly true that the doctrine that phenomenal unity is confined to phenomenal contents that are both momentary and simultaneous has a historical pedigree – it can be traced back to Kant, and endorsed by many of the nineenteenth century advocates of the specious present, such as Meinong and Brentano. However, actual arguments for *why* experiential unity is by its very nature confined to momentary states, and cannot extend any distance through time, are thin on the ground – in fact, I know of none. It could well be that philosophers who took this line did so because they were working on a (largely unvoiced) assumption that Presentism is the correct view of time – this was certainly the case for Brentano. As I have already noted, if reality is confined to a single strictly momentary present, our experiences of change and persistence must be too. In any event, for better or worse, we are no longer in the nineeteenth century, and Presentism is far from being the only metaphysically viable account of time on offer. Indeed, in the eyes of many metaphysicians, the block universe is the only conception of time which is not fatally flawed. And since all times and events in a block universe coexist (though of course, not simultaneously) it easily meets the requirements of the Extensionalist account of temporal experience. If everything in the universe, past, present and future, is fully and equally real, then experiences are too. In which case, there is no obstacle – posed by time, at any rate – to contents separated by a second or so being experienced together.

This is not to say that Extensionalism necessarily *requires* a block-type universe in order to be a viable option, there may be other options – I will be returning to this issue shortly. Before doing so I want spend a little time considering the respective merits of the Retentional and Extensional models of temporal experience.

5.4 Simplicity and Continuity

There is no denying that the Retentional model makes fewer metaphysical demands than its Extensional rival. The momentary states of consciousness in which, according to the Retentionalist, change-experience is packaged can exist equally well in a universe where only a momentary slice of reality is real as they can in a universe of the block variety. The Extensional model does not have so many degrees of freedom. But there are several other respects in which the Extensional model looks to have a clear advantage. I will focus here on just two.

Retentionalists are offering what is, in effect, a *two-dimensional* model of temporal experience. The change and persistence that we directly apprehend exist in episodes of experience which lack temporal extension in ordinary objective time. If the Retentional account is right, *at each instant* of ordinary time during our waking lives our consciousness has temporal depth, a temporal extension which does not exist *as* temporal extension in ordinary time.[7] The Extensional account, in

[7] Some leading Retentional theorists fully accept this characterization, e.g. Broad (1938).

contrast, is resolutely one-dimensional: the specious presents within which we experience change and persistence extend a short distance through ordinary objective time, just as they seem to.

There is a significant sense in which the Extensional account is the simpler of the two. This is because, quite simply, it offers an account of temporal experience which requires less exotic generative machinery than the Retentional alternative. It does not require a mechanism that is capable of creating a dimension of time – of duration – that exists (in effect) orthogonally to ordinary time. This alone would be a sufficient reason for preferring the Extensional theory if the two accounts were equally matched on all other fronts. In fact, I think that when it comes to accommodating temporal phenomenology the Extensional account has additional clear advantages. The problems for the Retentional model that I have in mind – and have explored in more detail elsewhere (e.g. Dainton 2006, 2011a, b, 2014a) – are quite simple, but serious and damaging, nonetheless.

Think of what it is like to hear a rapid succession of brief tones, e.g. the auditory experiences you would have if you heard C-D-E-F-G played on a harpsichord. What would your experience be like? Evidently, you would hear C being followed by D, D being followed by E, and similarly for the remainder of the sounds in the sequence. In experiencing this succession, there is a phenomenological feature that is perfectly obvious, but one that it is important not to overlook. Although you experience the successions C-D and D-E, you don't hear the D-tone twice over, you hear it just once. To put it another way, the D-tone which you experience following on from C is one and the same D-tone that you experience being following by E.

The Extensional theorist has no difficulty whatsoever in accommodating this very plausible description of what it is like to experience a succession of tones. To keep things as simple as possible, let's suppose that each of the tones in the C-D-E-F-G sequence is of the same duration, and this duration is half the duration of the specious present. For the Extensionalist, this short stretch of an auditory stream of consciousness has the structure pictured below.

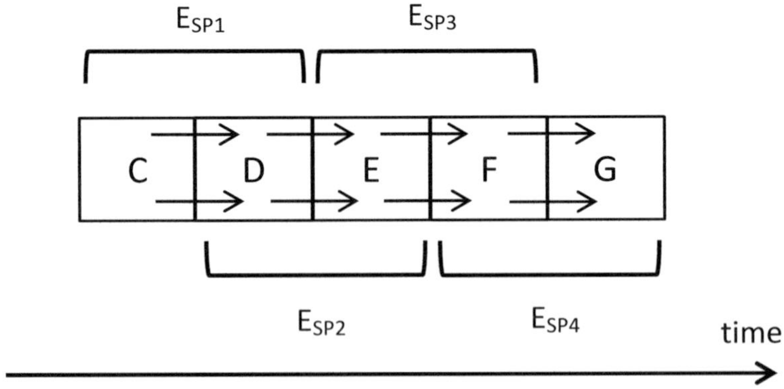

As can be seen, this stream-segment consists of four specious presents, E_{SP1} comprising the succession C-D, E_{SP2} comprising D-E, E_{SP3} comprising E-F and E_{SP4}, comprising F-G, with the interior arrows indicating the direction of experienced flow or passage. As can also be seen, these specious presents partially overlap, with E_{SP1} and E_{SP2} sharing a common part, in the form of the D-tone, which constitutes the second half of the former and the first half of the latter. The same holds *mutatis mutandis* for E_{SP2} and E_{SP3}, which share the E-tone, and E_{SP3} and E_{SP4}, which share the F-tone. By virtue of this partial overlap, the Extensionalist can easily explain how it is that we experience what we do, namely C flowing into D, and D (numerically the same D-tone) flowing into E, and similarly for the successions E-F and F-G.

The situation as seen from the Retentional perspective is depicted below. Once again we have four specious presents, but of the Retentional variety, i.e. momentary episodes of experience with dynamic contents.

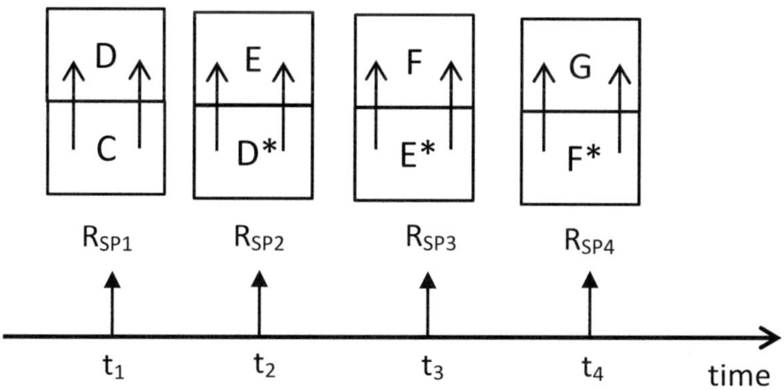

The content of R_{SP1} consists of C-being-followed-by-D, with this apparently extended (temporally) experience existing at a single instant of ordinary time t_1– indicated in the figure by its vertical orientation with respect to the horizontal – and the same holds for the experienced successions housed in R_{SP2}, R_{SP3} and R_{SP4}, which occur at t_2, t_3 and t_4 respectively. (It should now be clear, if it wasn't previously, why experienced change on the Retentional model exists "orthogonally" to ordinary time.) Since Retentional specious presents exist at different times, each consists of a numerically distinct episode of experience that is entirely self-contained. As is evident, specious presents in this guise do not overlap by virtue of possessing common parts in the manner of their Extensional counterparts. What they *do* possess, however, are parts that are qualitatively identical. For example, both R_{SP1} and R_{SP2} feature presentations (or re-presentations) of a D-tone, but since they occur at different times, the D-type experiencing in R_{SP1} is numerically distinct from its counterpart in R_{SP2}. To reflect this difference, the D-tone in R_{SP2} is marked with an "*", and similarly for the E* in R_{SP3} and F* in R_{SP4}.

There are various questions which one might pose regarding these two very different accounts of the temporal structure of our streams of consciousness, but let's focus here on just one issue: which best accommodates the phenomenology of experienced successions?[8] At first glance it might seem as though the difference in the experiential structures depicted above will make little difference on the phenomenological level. After all, in the case in question, we experience C-flowing-into-D, and D-flowing-into E (for example), and these experienced transitions are recognized by both theories: succession C-D is found in both E_{SP1} and R_{SP1}, whereas D-E is to be found in E_{SP2} and R_{SP2}. More generally, it looks as though both models accommodate all the experienced successions which the subject in question experiences during the period in question, albeit in very different ways.

In fact, the situation is by no means so straightforward. When we hear a three-phase succession such as C-D-E in our example, the succession is *identity-preserving* in this sense: we experience C flowing-into-D, and D-flowing into-E, where the "D" which is experienced as following on from C is the very same experience, numerically, as the "D" which is experienced as flowing into E. Call these "IP-successions". As I have already noted, the support for this claim is phenomenological. If asked to describe what it's like to experience C-D-E, it would – I think it safe to say – strike most of us as entirely natural to say "I heard C being followed by D, and D – the very same D – being followed by E". It would be odd in the extreme to say something along these lines: "I heard C being followed by D, and then D – a *completely different instance* of a D-tone – being followed by E". IP-successions play an absolutely crucial role in the continuity of our consciousness; if they didn't exist, our consciousness would not be continuous in the way it seems to be – and hence (very probably) *is*.

As should be clear, the Extensionalist has no trouble at all in accommodating IP-successions. For the Extensionalist, the sensory continuity we find in our typical streams of consciousness consists of overlap by part-sharing, and this form of overlap automatically delivers experienced successions of the IP-variety. Recall how the D experienced in the later phase of E_{SP1} is numerically identical with the D experienced in the initial phase of E_{SP2}; given this identity, *of course* we experience C being followed by D, and (the very same) D being followed by E. The Extensional approach is capable of explaining how, from moment to moment, our consciousness manages to possess the deep and distinctive form of continuity it seems to possess.

In very sharp contrast, the Retentional model cannot explain this at all, and proponents of this approach are obliged to adopt a quite revisionary account of what our experience is like. Retentionalists have no difficulty accommodating *some* experienced successions. In our simple example, successions C-D, E-F, and F-G are all handled in a straightforward fashion: these successions correspond to the dynamic contents of individual specious presents. However, when it comes to how

[8] One issue concerns the repetitions which are inherent in tale told by the Retentionalist. Since D (for example) is experienced in R_{SP1} and also in R_{SP2}, won't it be experienced twice over? In Dainton (2014a) I argue that this problem is less easily defused than Retentionalists often suppose.

these specious presents are related to their neighbours, serious problems soon emerge. Since individual specious presents are experientially self-contained, successions of the IP-variety simply cannot exist, or at least, not where these involve contents distributed among earlier and later specious presents. We can experience C-D and D-E, but we cannot experience C being followed by D, and (the very same) D being followed by E. Since IP-successions are such a ubiquitous feature of our ordinary streams of consciousness, this deficit is undeniably damaging for the Retentionalist. By virtue of enclosing our experience of change and persistence inside self-contained quasi-atomic episodes of experience, this model of temporal consciousness fails to do justice to the continuity which our streams of consciousness manifestly possess.

The encapsulation inherent in the Retentional approach has a second phenomenologically unrealistic consequence, one that is yet to be mentioned. Consider again our simple stream of auditory consciousness consisting of five successive notes.

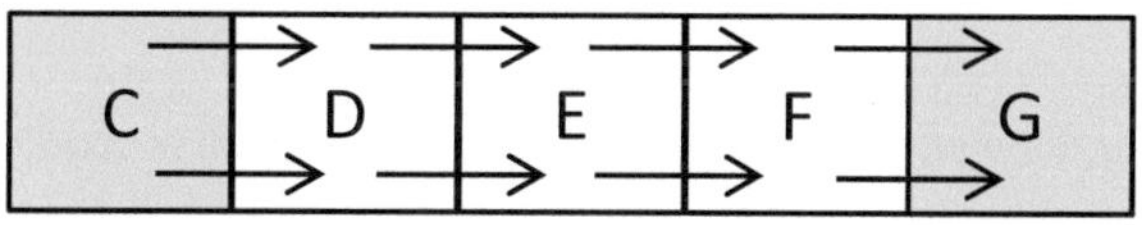

Typical notes, such as D, E and F, are experienced as *parts of successions.* The D-tone, for example, is heard as following-on-from C and being-followed-by E. It is very plausible to suppose that a D-tone that is experienced in the midst of such a sequence will differ, phenomenologically, from a D-tone which exists in utter and complete isolation experientially speaking, i.e. a D-tone that is not heard as following on from some other sound, or being followed by some other sound. Consequently, it is plausible to think there will be a noticeable difference in subjective character between these two cases.

The C- and G-tones – shaded in grey in the figure above, to mark them out – occurring as they do at either end of this brief stretch of auditory consciousness, are different again. The C-tone is not experienced as being preceded by another sound, but it *is* experienced as flowing into the D-tone; analogously, the G is experienced as following on from the F, but is *not* experienced as flowing into any other sound. Recalling the typographic nomenclature, we can call experiences such as C *phenomenal orphans* (they occur at the start of a succession, preceded by nothing) and experiences such as G *phenomenal widows* (they occur at the end of a succession, and are followed by nothing). Again, it is plausible to suppose that, in general, the subjective character of a phenomenal widow or orphan will noticeably different, even if only subtly, from experiences that do not have this status, experiences such as D, E and F which occur in the midst of IP-successions.

Bearing this in mind, let us return to considering the way in which the Extensional and Retentional theorists characterizes the C-D-E-F-G succession, and focus our attention on the distribution of phenomenal widows and orphans. The diagram below illustrates the situation as viewed from the Extensional perspective, and it looks entirely realistic, phenomenologically at least. Since D, E and F are all experienced in the midst of IP-successions – and hence experienced as following on from *and* being followed by some other auditory experience – the only phenomenal orphans and widows (shown shaded in grey) are the experiences of the C- and G- tones. This corresponds precisely with what one would expect.

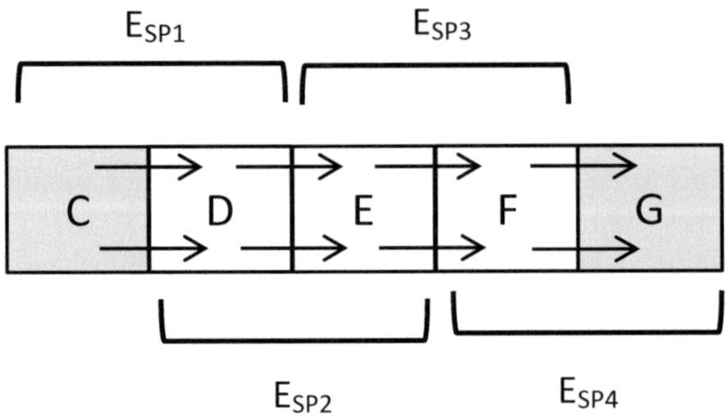

The situation viewed from the Retentional perspective could not be more different.

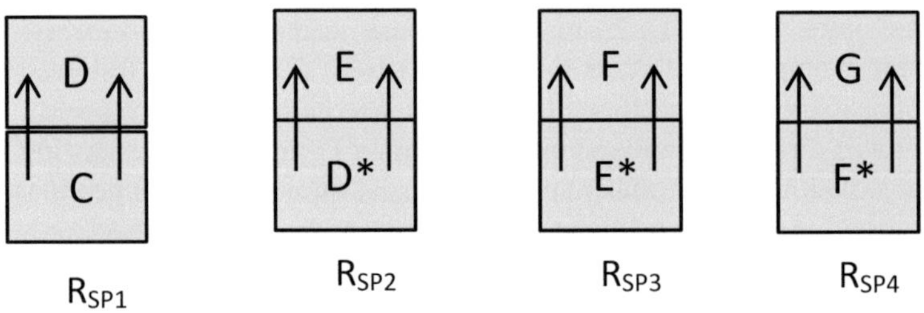

Here, each and every part of every specious present is shaded in grey. This is because each and every one of the depicted parts has now been transformed either into a phenomenal orphan or a phenomenal widow. The experiences corresponding to D, E, F and G are all experienced as *being followed by nothing*, whereas the experiences corresponding to C, D*, E* and F* are all experienced as *being preceded by nothing*. Once again, I suggest, this is not how our own experience seems when we hear a succession such as C-D-E-F-G. Our experience is fully continuous: we hear each brief phase, with the exceptions of the very first and the very last, as either being preceded by another tone, or followed by another tone. The Retentionalist introduces *discontinuities* in our experience that simply do not seem to exist. And when it comes to claims regarding the character of our experience, "how it seems" is of course what matters.

Summing up, the Retentionalist can offer us an account of what is going on within individual specious presents which is not obviously absurd or incoherent. True, it does require seemingly-extended dynamic phenomenal contents to exist in conscious states that are (objectively) momentary. But since we are sure of very little when it comes to the issue of what and how consciousness is produced in our universe, it is difficult – with any confidence – to deem this impossible. However, if we are seeking to understand temporal experience in all its forms and manifestations, we don't just need an account of individual specious presents, we also need an account of how distinct specious presents are inter-related, so as to form the longer streams of consciousness we all enjoy. As we have seen latterly, the experientially self-contained nature of Retentionalist specious presents makes this very much an uphill task. In contrast, by virtue of being temporally extended, there are possibilities open to specious presents of the Extensional variety that are closed off to the Retentionalist. In particular, the temporal extension that is the defining trait of Extensionalist specious presents permits them to partially overlap and share parts. This in turn permits them to combine to form extended streams of consciousness possessing the same continuous character as the streams we enjoy. So in the absence of further evidence or arguments, it looks very much as though we should reject the Retentional account and accept the Extensional alternative.

5.5 Phenomenal Passage: Unleashed and Unconfined

Smolin suggests that being *drenched* in time is "a fundamental attribute of conscious experience" (2013b: 32). The formulation strikes me as apt. What Smolin means by it is that our experience is *temporal* through and through, and by "temporal" he means (I take it) that all our experience, without exception, exhibits temporal *passage*, albeit of the phenomenal variety: it is everywhere *flowing*, and flowing in a particular direction: from the present, away into the past.

This form of passage takes many different forms. There is the distinctive dynamism of the direct experience of motion with which we started. There is the quite different, but no less dynamic – no less *flowing* – character of auditory experience. Think of what it's like to hear a succession of different musical notes, or a single note – played on a cello, say – continuing on and on. We don't just hear continuity-without-change, we can also see it: there is something distinctive that it is like to see (say) the blueness of the sky *continuing on* being blue. Our bodily sensations also exhibit a flowing character: think of what it is like to feel the hot sun warming one's skin, or (more obviously) water flowing through one's fingers – in a related vein, think of what it's like to chew on a morsel of food. Our inner consciousness is similarly dynamic, as one thought (or mental image, or memory) gives way to the next, then the next.

If we accept that phenomenal passage is a reality, we must accept the consequences.

I mentioned one of these consequences earlier. Even if we do live in a universe of the block variety, a universe where there is no moving metaphysically privileged present, a universe where all events are fully and equally real, it is wrong to conclude from this that our universe is entirely static or "frozen", as Slezak (following the lead of many others) describes it. This characterization is misleading because our experience is itself a part of the universe, and it is difficult to conceive of any experience existing which does not exhibit phenomenal passage, in some form or other.

As I also remarked earlier, the precise extent to which the dynamism and passage that exists in experience infects or pervades the universe as a whole depends on the precise relationship between experience, the properties we find in our experience, and the rest of reality. The different views on this issue have very different consequences.

Let's suppose, first of all, that the sensory qualities which constitute our sensory and perceptual experience exist only within our experiences. So although the redness of the rose, or the blueness of the sky, the pain in your ankle, or the noise made by a passing car all *seem* to be out there in the surrounding world, they are in reality only features of your experience, and they exist only within your consciousness. Although highly counterintuitive, at least on first acquaintance, this view of perception and perceptual qualities has been largely dominant since the days of Galileo, Descartes, Locke, and the Scientific Revolution. If this view *is* correct, then the dynamic features of our experience will also be confined only to our consciousness. They will not extend to the physical world external our consciousness.

The representational (or "indirect realist") view of perception may be the dominant view, but it is by no means the only view. There have always been defenders of direct (or naive) realism, and there continue to be. On this alternative view, perception is much as it seems: colours and sounds, at least in ordinary non-hallucinatory cases, are out there in the surrounding world, just as they appear to be. If *this* theory of perception is correct, the forms of passage we encounter in our ordinary experience are not confined to our consciousness, they are out there in the world and merely sensed by us in perceptual encounters. The consequences are profound. Since P-motion will not be confined to our experience, the highly dynamic *flickering* of a yellow flame will be a real feature of the flame as much as the *yellowness.* The same applies to the *rippling* of the blue waters of a pond, or the turbulent *roiling* invariably found on the yellow-orange surface of the sun (and other stars, needless to say). If direct realism is true, the physical world itself will be as dynamic as it is coloured.

If perceptual qualities such as colour are confined to our consciousness – currently a more popular view in philosophy and the cognitive sciences – the consequences of recognising that these qualities exhibit passage is diminished, but it by no means vanishes. But much now hangs on what the relationship is between consciousness and the rest of reality.

Suppose first of all that some form of dualism is true, and experience is non-physical in nature. If so, then the dynamism inherent in experience will not be a feature of the physical world, but it will still be fully be a part of the universe as a whole, the universe in the form of "the sum total of what exists". If this scenario obtains, then the *physical* world really would be static or frozen, even if reality as a whole is not. In saying this I am, of course, continuing to assume that the physical world is of the block variety. But the point remains valid – in one important sense at least – even if some form of metaphysical passage (as we can call it) obtains. If the physical world takes the form of a growing block, for example, if dualism is true it remains the case that the physical realm is entirely devoid of *phenomenal* passage, and hence lacks the form of temporal flow or dynamism with which we are most directly and intimately acquainted.[9]

But dualism is by no means the only viable-looking solution to the so-called "hard problem", and it is not necessarily the most promising-looking. What makes the problem of consciousness so hard is the difficulty of finding a place for phenomenal properties such as colour, sound and felt warmth (or pain) any place in the physical world. This difficulty derives, ultimately, from the conception of the basic nature of the physical world bequeathed to us by the Scientific Revolution. If the fundamental properties of elementary physical things are confined to those recognized by physics – size, motion, mass, charge, energy, spin, and so on – then there are no phenomenal properties whatsoever at the base-level of reality. Elementary particles don't possess intrinsic colour properties, and it is very hard to see how they could come to possess a new intrinsic property – such as colour of the phenomenal variety – merely by being put into a different spatial arrangement, or engaging in the sorts of interactions which physics allows. Or more bluntly, it is hard to see how configuring elementary particles into living human brains could ever result in them possessing entirely new intrinsic properties, of a kind fundamentally different to any they previously possessed – not without the intervention of magic or miracles. But unless this is the case, the phenomenal properties we encounter in our experience cannot be produced in or by our brains, for our brains are composed entirely of elementary physical ingredients.[10]

One familiar solution to this problem is to opt for some form of dualism, and accept that phenomenal properties don't exist anywhere in the physical realm. Another, and in many ways more appealing option, is to hold that consciousness is wholly physical, and wholly a product of physical processes in our brains. As for how this is possible, the way forward is to hold that some physical things or processes possess intrinsic properties over and above those recognised by current physics. If these additional properties are *phenomenal* in nature, the previously insurmountable obstacle to taking experiences themselves to be physical in nature is no longer an obstacle at all, at least in principle. The basic materials, as it were, for producing experience are now to be found in the physical realm.

This position, which has recently been known as "Russellian Monism", can itself be developed in different ways. One possibility is that phenomenal properties – and hence conscious states – are to be found in a narrow range of physical phenomena. For example, perhaps within a particular form of field generated within human brains. If so, the dynamism characteristic of experience is itself confined to a very restricted portion of physical reality. But there are other possibilities. Many Russellians argue that it is more plausible to suppose that phenomenal properties

---

[9] For more on the distinction between metaphysical and phenomenal forms of passage, and reasons for supposing that the latter is largely independent of the former, see Dainton (2011a, b, 2014a); for more on the metaphysical implications of phenomenal passage see Paul (2015).

[10] This line of argument is expounded more fully in Dainton (2014b, Chap. 8).

must reside within *all* elementary physical things: particles, fields, and space-time itself (assuming it exists as a substantial entity in its own right). This position will appeal to anyone who is inclined to think that the physical realm is fundamentally homogeneous. If everything that is physical has the same general kind of intrinsic nature, then all physical things will have an intrinsic character of an experiential kind if some do. The fact that all physical things are interconvertible with energy may not conclusively prove that the homogeneity thesis is true, but it certainly provides it with some rationale.

Russellianism in this guise is indistinguishable from *panpsychism*, the venerable metaphysical doctrine according to which every last thing in the physical realm, from the smallest to the largest, has some degree of conscious mentality. The consequences are once again dramatic. Most panpsychists willingly concede that the consciousness of elementary particles such as neutrinos, quarks and electrons (or superstrings, for that matter) is likely to be very simple in character, and that complex forms of consciousness, such as we enjoy, only come into existence when matter is organised into more complex systems, such as human brains. Even so, it is difficult to conceive of even a simple form of consciousness that is entirely lacking in dynamic temporal characteristics, which does not involve experienced change or persistence, and hence phenomenal passage. If this is the case then passage of the phenomenal kind will be utterly ubiquitous in our universe – even if our universe turns out to be of the block variety.

We should not get ahead of ourselves. The case for taking panpsychism seriously is more powerful than is often thought, but the doctrine faces severe problems.[11] These problems could easily turn out to be insuperable, in which case we will have to look elsewhere for a solution to the hard problem. More generally, the debates concerning the relationship between the experiential and the physical realms are not likely to be brought to a definitive conclusion any time soon. However, as I hope is also clear, although we cannot yet know how these debates will be resolved, their resolutions have very different, and in some cases quite radical implications for the extent to which our universe possesses an inherent dynamic character.

5.6 Presentism and Experience

The fact that our experience is intrinsically dynamic has, as we have just seen, potentially significant implication for the extent to which our universe is intrinsically dynamic. The temporality of experience has cosmological implications of a quite different kind.

We saw earlier that the Extensional model of temporal experience looks to be the most promising of the alternatives on offer. For in sharp contrast with the competing Retentional account, it does not render our streams of consciousness into succes-

[11] See Strawson (2008, Chaps. 1–2) for a vigourous defense of panpsychism, and Chalmers (forthcoming, 2015) for an assessment of the difficulties.

sions of disconnected fragments, and so can do full justice to the phenomenological evidence. Our experience seems deeply continuous, with each brief phase flowing into the next (in an identity-preserving way), and the overlap structure posited by the Extensionalist can account for this in a very natural manner.

However, as we also saw, the Extensional model may well have broader implications for the metaphysical nature of time. Since it packages specious presents into episodes of experiencing that are, objectively, momentary (or very close to it), the Retentional model is fully compatible with all the leading views concerning the nature of time. Specious presents of the kind the Retentional theorist posits can exist in a block universe, but they can also exist in growing block and Presentist universes. The Extensional model is not so accommodating. Since the specious presents it posits extend a short way through ordinary objective time – a second or so, say – these unified units of experience make requirements on time itself. They can exist perfectly well in block and growing block universes, since experiences of the required duration can exist in both types of world. But they cannot exist if Presentism is true, and reality in its entirety is confined to a single momentary phase. There just *isn't time* in a Presentist universe for specious presents of the Extensional variety to exist.

This incompatibility could be dismissed as of little importance if Presentism itself could be dismissed as a metaphysical *curio* that had been shown to be untenable by advances in science such as Einstein's special theory of relativity. But as we saw at the outset, the "block consensus" among physicists and cosmologists is showing signs of breaking down. Consequently, as things currently stand, we cannot dismiss Presentism out of hand simply by appealing to scientific considerations.

So far as Extensional specious presents are concerned, a Presentist might respond thus: "Yes, it's true that I believe that concrete reality consists of a single momentary present, but I also believe there are *truths* about the past, and in some cases about future too. In which case, there can be truths about what I have been *experiencing* for the past few seconds, and perhaps for what I will be experiencing over the next few seconds. Isn't this enough to meet the demands of the Extensional account? How could it fail to be?"

In fact, it could fail quite easily. Many Presentists do indeed want to make room for truths referring to past and future happenings. However, when it comes to truth-makers for statements about non-present events, they have to turn to something other than past and future events themselves (which of course, don't exist on their view). Currently the most popular strategy is to appeal to abstract objects, in the form of propositions, to serve as the required truth-makers. In rough outline, the idea runs thus. There is (we might suppose) a set of propositions P which fully describes, down to the last detail, the entire history of the universe up until the present. If so, then a statement about some past event occurring at time t is true if it makes the same claim about reality as some proposition in set P.

This might provide (after a fashion) a vast realm of past truth, but it is very difficult to see that it could provide the Extensionalist with what they need.

From the Extensional perspective, a specious present is a stretch of unified experience which extends a short way though time. If, in a simple case, the content of a particular specious present is an auditory succession [C-D], then the C- and D-tones will be experienced together-in-succession (rather than simultaneously). If the C- and D-tones are experienced in this way, if they are diachronically co-conscious, the experiencing of the C and the experiencing of the D must both be real; they must co-exist, albeit not at the same time. More generally, any unified experience E, if E has parts, and these parts are co-conscious, synchronically or diachronically, these parts *must* all be real. A unified experiential whole cannot consist of a combination of parts, some of which exist and some of which don't.

Hence the problem for the Presentist. If we suppose Presentism is true, then when D is being experienced, the experiencing of C does not exist, since it lies in the past. If the experiencing of C is wholly unreal, then the Extensional specious present consisting of [C-D] cannot exist either. The Presentist will point out that the non-existence of a given experience does not mean there are no *facts* about that experience. But this helps not at all. For according to the Presentist's way of thinking, what *makes it true* that your experience of C occurred just prior to your experience of D is an abstract object, a proposition. This falls short of what is required. For the Extensionalist, the specious present [C-D] is a unified experience possessing *experiences* as parts. The required experiences simply aren't available if Presentism is true. By the time D occurs, the experience of C has vanished, and all that remains is the proposition describing C's occurrence. A conscious state cannot be made of materials such as these.[12]

If Presentism is incompatible with the Extensional view of temporal experience, then we look to have good reason for regarding Presentism as untenable. It is very hard to deny that our experience *is* temporal: we directly apprehend change, succession and persistence, and we do so within streams of consciousness which are profoundly continuous. Since of the available alternatives, only the Extensional model can accommodate these hard-to-deny claims about our experience, we have every reason to accept this model, and reject Presentism.

However, the situation is by no means so clear-cut. I have been assuming thus far that Presentists are committed to the claim that reality is confined to a *single momentary* phase. This is the standard way of construing Presentism, but it is not the only possible way. It is also possible to hold that the present is *more* than strictly momentary, that it has some temporal extension. For obvious reasons, let's call this *Extensional Presentism.*

A universe of this general type can take very different forms. We'll begin with the most straightforward. To arrive at it, take a standard block universe, and imagine it neatly sliced into a series of very thin blocks – to keep things simple, we'll suppose that these blocks all have a duration of precisely one second. Then imagine these thin slices of reality coming into existence one by one, and then ceasing to exist (see the figure below).

[12] See Dainton (2010b) for a more detailed exposition of this line of argument.

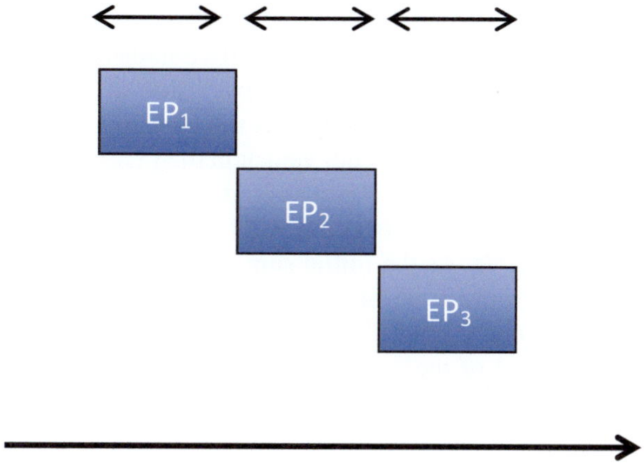

To register their distinctive (non-momentary) character, we can refer to these transient blocks, only one of which exists at any given time, as "thick" or "extended" presents.

This model of time goes part-way towards meeting the demands of Extensional model, but it does not go far enough. On the plus side, there is now enough temporal room in the universe for Extensional specious presents to exist. If, as we have been supposing, a typical specious presents lasts for a second (or a little less), then each of EP_1, EP_2 and EP_3 could house a single specious present, since each of these brief blocks of reality also lasts for a second. But a universe of this kind is clearly incapable of accommodating the *continuity* we find in our streams of consciousness. Since EP_1 has suffered annihilation by the time EP_2 enters the picture, the later parts of the former cannot be experienced as flowing into the earlier parts of the latter, so there is no possibility of a stream of consciousness continuing on, without interruption, from EP_1 to EP_2. In short, if the universe were like this, then our streams of consciousness would be confined to brief discrete pulses, with the catastrophic results we noted earlier.

But a Presentist universe needn't take the form of a succession of entirely discrete blocks of reality. There is no reason why it can't take the form of a succession of *partially overlapping* blocks, in the form depicted below.

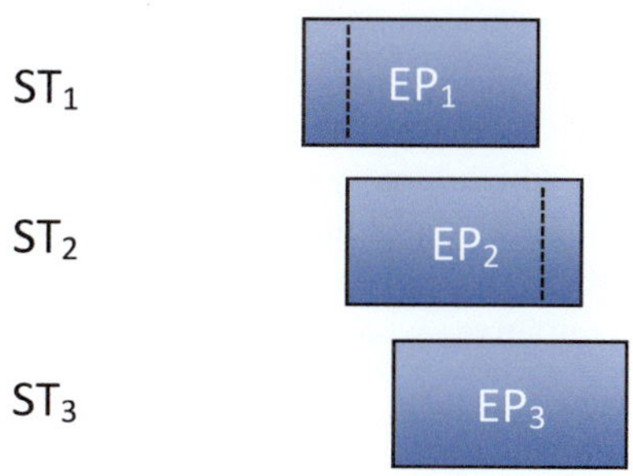

Here, EP_1, EP_2 and EP_3 are again extended presents, or brief blocks of reality. But this time, they are not entirely distinct and self-contained. Instead, each largely overlaps with its successor and predecessor, by virtue of sharing (numerically identical) parts. Temporal passage exists in this universe too, but it is a more modest affair. Entire extended presents do not vanish in their entirety, to be replaced by entirely new and distinct presents. Passage consists of no more than very brief temporal *slices* of extended presents being annihilated or created. The process is depicted in the figure above. As the initial sum total of reality, ST_1, gives way to the next, ST_2, the earliest phase of EP_1 is annihilated (the part lying to the left of the dotted line), while a new brief phase is created at the leading edge of reality (the part lying to the right of the dotted line EP_2). As time continues to pass, the earliest phase of EP_2 is annihilated and a new phase is created, to yield ST_3. And so it goes on.

This *Overlap Presentism* (as I will call it) can itself take different forms. I have been assuming that the temporal breadth of an extended present is one second, but there might be universes where it is half a second, a hundredth of a second, or an hour. Similarly, the magnitude of the increments and annihilations which constitute temporal passage could vary from universe to universe. In some worlds these might be of the order of the Planck duration, in others they might be a billionth or a millionth of a second – there are many possibilities. For present purposes, what matters is the *general shape* of the account. By combining an extended present with overlap between successive universe phases, the Presentist's position is rendered more plausible – or at least, less problematic – in two important respects.

First, many of the disadvantages of the discrete non-overlapping block model are avoided. If successive (extended) presents *don't* overlap by part-sharing, each is – in effect – an entirely separate short-lived universe. This radical isolation leads to obvious difficulties. It is not easy, for example, to see how the events in distinct presents can causally interact with one another. More generally, is also difficult to see how one phase of the universe can *produce* the next, if the earlier has utterly ceased to exist by the time the next comes into existence. How can non-existent objects or events *produce* anything? Aren't non-existent things causally inert? These problems are entirely avoided by adopting the overlap model, where causes and effects can co-exist within the confines of a single (brief) reality-block.[13]

Second, and more relevantly to our main concerns here, the conception of time that the Overlap Presentist proposes is entirely compatible with the Extensional view of temporal experience. In fact, the two fit like hand and glove. According to the Extensionalist, a stream of consciousness consists of a succession of temporally extended specious presents which largely overlap by part-sharing. According to the Overlap Presentist, reality as a whole has precisely the same structure. If reality does have this structure, there is no obstacle to our streams of consciousness being fully continuous, even in universes where Presentism (in one form at least) obtains.

---

[13] I explore the overlap version of Presentism in more detail in Dainton (2010b: Chap. 6), where I call it "Compound Presentism".

At this juncture an objection along the following lines might be occurring to some:

> The structural similarities are undeniable. But the precise fit between Overlap Presentism and Extensionalism is largely an artifact of the way you have presented the two accounts. You've deliberately engineered things so they fit perfectly: you worked on the assumption that Extensional specious present are of precisely the same duration as Extended presents, one second. Why assume this? Couldn't an Extended present be longer than a second – longer than specious present, however long that turns out to be?

In one sense this is perfectly true. I have been working on the assumption that the specious present's apparent duration is around one second, and I did assume – at least initially – that the Extended present is of around the same duration. That said, the first assumption is phenomenologically plausible (or at least, it seems so to me), and although findings from the cognitive sciences might well result in this approximation being refined – e.g. it may turn out that the specious present varies from person to person, or from one occasion to another – I would be surprised if it were revised in a radical way.

The second assumption also has a rationale. If Presentism is an intuitively appealing view of time it is because it fits so well with our *experience* of time. If the past exists, we are not aware of it – we don't directly experience it in the way we experience the (extended) present – and similarly for the future. Our immediate experience is confined to the *now*. Since the principle driving force behind Presentism is to confine reality to what is experienced, if it turns out that the experienced present has some duration, then so too does (or should) the Presentist's present. Hence the notion that the experiential present could have a different duration from the Presentist's temporal present runs counter to the primary rationale for taking Presentism seriously in the first place.

In a more speculative vein, it is worth mentioning another possibility, another reason why the durations of the Extended and specious presents might coincide.

The contents of our conscious states at any given time are *co-conscious*, they are all experienced together. This is a deep and distinctive mode of unity, and down the ages philosophers have striven to explain it in ways which register its distinctiveness. In the *Theaetetus*, after drawing our attention to the fact that our ears cannot perceive what we can see, and our eyes cannot perceive what our ears can hear, Plato posited a higher-order, "common sense" to explain the unity of consciousness. If auditory and visual contents are experienced together, it is because they fall under this multi-modality mode sensing. Descartes took a different route. Our diverse forms of experience manage to be experienced together because they are states of a metaphysically primitive substance, i.e. one lacking constituent parts, whose essence is to *be* conscious.

There is, however, a way of registering the distinctive depth of phenomenal unity that is less metaphysically laden. It is plausible to think that the co-consciousness relationship is necessarily *transitive*. That is to say, if one experienced item X is co-conscious with another experienced item Y, and Y is co-conscious with Z, then X and Z must also be co-conscious. It is certainly very hard to imagine a state of affairs in which X and Y are experienced together, and Y and Z are experienced together, but X and Z remain wholly isolated from one another, phenomenologically. As for why this should be the case, one explanation is straightforward: if two contents are connected by the co-consciousness relationship they are so *closely* bound together that it is impossible for some third content to be co-conscious with one without also being co-conscious with the other.

There is more to be said on this issue,[14] but there is at least a *prima facie* case for taking synchronic co-consciousness to be transitive. The situation in the diachronic case is quite different. If the auditory succession [C-D] constitutes a single specious present, then every part of this experience is co-conscious, synchronically or diachronically, with every other part. But over slightly longer intervals, the transitivity of diachronic co-consciousness breaks down. If you hear C-being-followed-by-D, and then hear D-being-followed-by E, your stream of consciousness during this period of time has this form: [C-D], [D-E]. In other words, C and D are diachronically co-conscious, as are D and E, but C and E are *not* diachronically co-conscious, since they occur in distinct (but partially overlapping) specious presents.

As for what we are to make of this, one possibility is that synchronic and diachronic co-consciousness are simply different when it comes to transitivity. Synchronic transitivity is by its very nature transitive, but diachronic co-consciousness is not. Anyone who believes that their own streams of consciousness exist in a block universe has no alternative to taking this view. But if our universe has the sort of structure that is endorsed by Overlap Presentists, a different option opens up. The Presentist can maintain that co-consciousness in both its synchronic and diachronic forms *is* transitive, and necessarily so. The only reason why transitivity fails in the diachronic case is that the passage of time brings with it the annihilation of earlier phases of experience, as new phases are created. In the case of [C-D], [D-E], the reason why C isn't co-conscious with D is simply that C no longer exists – is no longer part of the sum total of reality – when E exists. If it *weren't* for the destructive agency of temporal passage C *would* be co-conscious with E.[15]

5.7 Bolzmann Brains

I have been arguing that the Extensional model of temporal experience has cosmological implications of a non-negligible kind: at least two forms of presentism can be ruled out, but one form very definitely cannot. Before drawing to a close I want to pause briefly to draw attention to an epistemological consequence.

---

[14] See Dainton 2006, Chap. 4, for further discussion.

[15] There is a further point to note in this connection. Anyone who is convinced that co-consciousness in all its forms is necessarily transitive can argue that Overlap Presentism *must* be true, since if we lived in a block universe all the experiences in our streams of consciousness – even those lasting for many hours – would all be co-conscious. And of course, this is very definitely not the case.

We start with an episode in the early history of theorizing about time-consciousness. In developing his views on these issues Brentano was both uncommonly clear-headed, and also influential. The influence stems from the fact that Husserl attended his lectures on the topic in 1880s, and developed his own – highly influential – views in response to his teacher's. Brentano was clear-headed because he saw, very early on, that one's views relating to the nature of time are not independent of the account of temporal experience that one can coherently adopt. Brentano was a committed presentist: he held that reality was confined to a momentary, durationless interface between past and future. He also believed that we directly experience change and persistence, which means that our consciousness must – in some way – possess temporal extension. Brentano reconciled these views by adopting a Retentional model of temporal experience: our consciousness is in fact confined to a single momentary phase, but possesses apparent temporal depth by virtue of the fact that our momentary sensory experience are accompanied by re-presentations of the experiences we had a few moments before – experiences which are no longer in existence.[16]

Brentano fully appreciated that the confinement of consciousness to momentary phases would have phenomenological consequences. In particular, the total lack of experiential connections between adjoining specious presents means our streams of consciousness are composed of a series of entirely self-contained episodes of experience. As Brentano also realized, one consequence of this is that our experience at any one time would be entirely unaffected if the past and future did not exist at all. In other words, if you only came into existence this very moment, and will cease an instant later, your experience could be – *would* be – just as it is. The streams of consciousness we all enjoy are indistinguishable in character from those of a being who exists only for an instant (or one who is continually being annihilated and then re-created *ex nihilo*).

Although Husserl found much to admire in Brentano's work on time-consciousness, he found this doctrine implausible in the extreme. For Husserl, the phenomenological evidence was unequivocal. Our consciousness is *not* confined to momentary self-contained, entirely isolated fragments; it exhibits a deep continuity: we our aware of each successive phase of our streams of consciousness flowing smoothly, without interruption, into the next. And because of this, we are aware – or can be – of our *consciousness itself* continuing to flow through extended periods of time. However, when Husserl tried to develop a theory of time-consciousness which renders it possible for us to be aware of both temporally extended phenomena *and* the continuity of own experiencing through time, he ran into serious problems. The Retentional framework he took over from Brentano makes it very difficult to avoid confining consciousness to experientially isolated self-contained episodes. Husserl struggled with the problem for decades, and was eventually driven to endorse an "absolute time-constituting consciousness" whose workings are elusive at best, and obscure at worst.

[16] For more on Brentano's evolving views on time-consciousness see Chisholm (1981).

In contrast, as we saw earlier, the Extensional model accommodates the continuity of our experience in a very natural way. It also provides us with a clear account of how our own experience differs from that of a momentary (or very short-lived) subject. So far as our actual streams of consciousness are concerned, each phase is connected to their successors and predecessors by diachronic co-consciousness; each phase is directly experienced as flowing into the next. The consciousness of a momentary or short-lived subject is very different. Such a subject's experiences are without exception *experientially disconnected* from any experiences which occur earlier or later. Since each phase of our own consciousness *is* experientially connected to earlier and later phases, we have empirical evidence, of the strongest sort, that we ourselves are not momentary subjects. If we were, our consciousness could not have the character it in fact has.

These considerations are very much relevant to some contemporary cosmological debates. Thanks to a string of recent developments, it now appears that our universe may well be *far bigger* than cosmologists had previously supposed, both temporally and spatially. Thanks to the recently confirmed existence of dark energy, it looks likely that the universe will keep on expanding forever, at an increasingly fast rate. Evidence deriving from analyses of the cosmic background radiation that the universe is (globally speaking) very close to flat has lent support to inflationary universe scenarios, which predict precisely that. What's more, the most promising inflationary theories are *eternal* in character. On this view, inflation is an endless process whereby quantum fluctuations in tiny regions of space give rise to universes of sizes comparable to our own, not once but over and over again, for ever and ever. The cosmos we are familiar with – the cosmos we investigate using our biggest telescopes – is vast almost beyond comprehending, but if eternal inflation exists, it is no more than a tiny "bubble universe" in a cosmos that is vaster by far – a cosmos that is quite possibly infinite in size.

Findings which suggest that our entire cosmos may well be infinite, both spatially and temporally, will no doubt be welcomed by those who enjoy contemplating endless vastnesses. But if the universe *is* this vast, there may be some rather disturbing implications, as Albrecht and Sorbo pointed out in a 2004 paper:

> A century ago Boltzmann considered a "cosmology" where the observed universe should be regarded as a rare fluctuation out of some equilibrium state. ...
>
> From this point of view, it is very surprising that we find the universe around us in such a low entropy state. In fact, the logical conclusion of this line of reasoning is utterly solipsistic. The most likely fluctuation consistent with everything you know is simply your brain (complete with "memories" of the Deep Hubble fields, WMAP date, etc.) fluctuating briefly out of chaos and then immediately equilibrating back into chaos again. This is sometimes called the "Boltzmann's Brain" paradox.

Boltzmann realized that although it is not very likely (at all!) that a system as vast, complex and ordered as the visible cosmos should be the product of nothing more than random interactions among atoms jostling one another in empty space, it is *possible* that purely random processes could create an ordered cosmos such as ours. Moreover, if the universe exists for an infinite period of time, it is *inevitable* that systems resembling the visible cosmos will arise, from time to time, even if most of the universe, for most of the time, consists of a thin soup of particles in a highly *disorganized* (high-entropy) state. However, what Boltzmann may not have appreciated is that systems that are far smaller and less complex than the entire known cosmos will also fluctuate into existence, and do so with far greater frequency. Lone stars or planets, an oak tree, a lump of granite, an ostrich egg – such objects are all simpler (and hence more probable) than an entire cosmos, and all could be produced in considerable numbers by random fluctuations. More worryingly, so too could momentary configurations of particles that (briefly) constitute healthy and functioning human brains. These hypothetical entities have become known as *Boltzmann brains*. As Sean Carroll puts it, in an eternal universe full of randomly interacting particles,

> No matter how many ordinary observers exist in our universe today, they would be dwarfed by the total number of Boltzmann brains to come. Any given observer is a collection of particles in some particular state, and that state will occur infinitely often, and the number of times it will be surrounded by high-entropy chaos will be enormously higher than the number of times it will arise as part of an "ordinary" universe. (2010: 233)

If you think that you can be pretty sure *you* are not one of these Boltzmann brains, then you haven't understood what's being proposed. The brains in question may be the product of a random (and incredibly improbable) particle fluctuation, but they are – or so we are to suppose – intrinsic duplicates of ordinary and fully-functional human brains, down to the last detail. Consequently, during their fleeting existence they enjoy conscious states of just the kind *we* enjoy, from moment to moment.[17] If Boltzmann brains can have the same kinds of experience as you or I, and brains of this sort *vastly* outnumber ordinary human beings, then the conclusion is obvious. If we live in an infinite universe, statistically it is far more likely than not that you and I *are* Boltzmann brains. And hence doomed to have a very brief existence!

However, all may not be lost. At least, not if the conclusions I defended earlier concerning the nature of temporal experience and the structure of our streams of consciousness are along the right lines.

On learning of the Boltzmann brain scenario, Brentano might very well have drawn the conclusion that it is highly probable that his own consciousness resides in a such a brain. If our streams of consciousness consist of successions of momentary specious presents that are discrete and entirely self-contained, in the way Brentano believed, then there is no (obvious) reason why the experience that could be had by the typical Boltzmann brain need be any different in character from our own experience. However, as we have seen, there are compelling reasons, of a phenomenological kind, for supposing that Brentano's atomistic account of the structure and

[17] The typical Boltzmann brain will find itself amid "high-entropy chaos", as Carroll puts it, i.e. in a universe where there are few, if any stars or other complex material objects in its vicinity. As a consequence, they will not be able to *perceive* a cosmos such as ours. But this does not matter. Since Boltzmann brains can easily suffer *hallucinations* which are radically misleading with regard to the real character of their surroundings, there might be many (many) such brains hallucinating – or dreaming – star-filled skies and Earth-like surroundings.

composition our streams of consciousness is mistaken. As Husserl appreciated fully, our consciousness does not consist of experientially disconnected brief pulses, it is deeply continuous, with each brief phase flowing seamless into the next. Consequently, if your experience *were* being produced by a brain that briefly fluctuates into existence, and lasts a short while – a second or so, but no longer – before vanishing again into chaos, then your experience would not and could not have the character it actually has.

To produce experiences of the kind you have been enjoying for the past few hours, a Boltzmann brain would have to exist continuously, and exactly resemble your brain in all experience-producing respects throughout this period. Needless to say, while the spontaneous formation of such a long-lived brain (and the life-support devices needed to keep it functioning!) is not logically impossible, it is immensely less likely than the spontaneous emergence of a brain which lasts only a second or so. The Boltzmann brain scenario cannot, perhaps, be dismissed altogether. But these basic phenomenological considerations point to its being considerably less of a threat to us than it might otherwise be.[18]

References

Albrecht, A., and L. Sorbo. 2004. Can the universe afford inflation? *Physical Review D*, http://arxiv.org/abs/hep-th/0405270

Bergson, H. 1911, *Creative Evolution*. Trans. A. Mitchell. New York: Henry Holt.

Broad, C.D. 1923. *Scientific thought*. London: Routledge and Kegan Paul.

Broad, C.D. 1938. A reply to my critics. In *The philosophy of C.D. Broad*, ed. P.A. Schilpp. New York: Tudor.

Carroll, S. 2010. *From here to eternity: The quest for the ultimate theory of time*. New York: Penguin.

Chalmers, D. 2015. Panpsychism and panprotopsychism. In T. Alter and Y. Nagasawa (eds.) Consciousness in the Physical World: Essays on Russellian Monism, Oxford University Press.

Chalmers, D. forthcoming. The combination problem for panpsychism. In *Panpsychism*, eds. G. Bruntrup and L. Jaskolla. Oxford University Press.

Chisholm, R. 1981. Brentano's analysis of the consciousness of time. *Midwest Studies in Philosophy* 6(1): 3–16.

Dainton, B. 2006. *Stream of consciousness*, 2nd ed. London: Routledge.

Dainton, B. 2008. Sensing change. *Philosophical Issues* 18(1): 362–384.

Dainton, B. 2010a. Temporal Consciousness. In *Stanford encyclopedia of philosophy*, http://plato.stanford.edu/entries/consciousness-temporal/

Dainton, B. 2010b. *Time and space*, 2nd ed. Durham: Acumen.

Dainton, B. 2011a. Time and temporal experience. In *The future of the philosophy of time*, ed. A. Bardon. New York: Routledge.

Dainton, B. 2011b. Time, passage and immediate experience. In *Oxford handbook of philosophy of time*. Oxford: Oxford University Press.

Dainton, B. 2014a. Flow, repetition and symmetry. In *Debates in the philosophy of time*, ed. N. Oaklander. London: Routledge.

Dainton, B. 2014b. *Self*. London: Penguin.

[18] For comments on earlier drafts my thanks to the Editors, Philip Goff and Galen Strawson.

Dolev, Y. 2014. Motion and passage: The old B-theory and phenomenology. In *Debates in the philosophy of time*, ed. N. Oaklander. London: Routledge.
Ellis, G. 2006. Physics in the real universe: Time and spacetime, http://arxiv.org/pdf/gr-qc/0605049.pdf
Ellis, G., and T. Rothman. 2010. Time and spacetime: The crystallizing block universe. *International Journal of Theoretical Physics* 49: 988–1003.
Paul, L. 2015. Experience and the Arrow. In A. Wilson (ed.) *Chance and Temporal Asymmetry*. Oxford: Oxford University Press.
Russell, B. 1903. *The principles of mathematics*. London: George Allen & Unwin.
Russell, B. 1917. Mathematics and metaphysicians. In *Mysticism and logic*. London: George Allen and Unwin.
Slezak, M. 2013. Saving time: Physics killed it. Do we need it back? *New Scientist*, 01 Nov
Smolin, L. 2013a. *Time reborn: From the crisis in physics to the future of the universe*. London: Allen Lane.
Smolin, L. 2013b. Temporal naturalism, http://arxiv.org/abs/1310.8539
Stern L. 1897/2005. Mental Presence-Time. Trans. N. De Warren. In *The new yearbook for phenomenology and phenomenological research*, ed. C. Wolfe, 205–216. London: College Publications.
Strawson, G. 2008. *Real materialism and other essays*. Oxford: Oxford University Press.

Chapter 6
From Physical Time to Human Time

Jenann Ismael

Abstract Time as experienced is said to have several properties that the physical image of time lacks.

In this paper, I outline a strategy for bridging the gap between the time of everyday experience and the time of physics that treats the Block Universe as a non-perspectival view of History and shows how to recover the everyday experience of time as a view of History through the eyes of the embedded, embodied participant in it. I also address questions about whether features of our temporal experience like passage and flow are properly thought of as illusory, the temptation to reify these features in the absolute fabric of the universe, and the question of whether this strategy takes passage seriously.

Keywords Temporal experience • Perspective • Passage • Flow • The openness of the future • Relativity • Block universe • *Sub specie aeternitatis* • McTaggart

Physics has forced us to revise our world-views in surprising ways and has also opened up new mysteries. The mysteries that get the most play outside of science are the mysteries at the frontiers of the physics of the very large and the very small. Almost everybody in the academy these days knows about quantum mechanics and the Higgs boson. Some may even know about Bell's Theorem. Everybody knows about black holes and dark matter. Some may even know about string theory and loop quantum gravity. These are the problems that tend to grab the popular imagination and also attract the attention of philosophers of physics. But some of the most difficult unsolved problems are much closer to the human scale and have to do with reconciling the way that physics tells us the universe *is* with the way that we experience it. So, for example, we do not have a good understanding of why time seems to have a direction, why the future seems different than the past, why time seems to flow, or even what this last thing means.

J. Ismael (✉)
Department of Philosophy, University of Arizona, Tucson, AZ 85721, USA
e-mail: jtismael@email.arizona.edu

Y. Dolev, M. Roubach (eds.), *Cosmological and Psychological Time*, Boston Studies in the Philosophy and History of Science 285,
DOI 10.1007/978-3-319-22590-6_6

Russell in the chapter entitled "The Abstractness of Physics" from *The Analysis of Matter*, remarks on the distance that has grown between common sense and physics:

> From [its] happy familiarity with the everyday world physics has been gradually driven by its own triumphs, like a monarch who has grown too grand to converse with his subjects…

In another passage he likens physics and perception to a pair of friends walking in conversation along opposite sides of a stream who don't notice as the stream gradually widens into a river so deep that they can no longer hear one another or get across.[1] The traditional approach to reconciling the manifest image[2] of the world with the image presented by physics is to focus on the logical relationship between macroscopic and microscopic descriptions of the world. At first, high-level structures like objects were assumed to be collections ('mereological sums', in the lingo) of low-level objects in a given arrangement.[3] That idea turned out to be a little too simple because the singular terms we apply at higher levels refer mostly to things whose criteria of identity over time are not the criteria of identity for collections of micro-level constituents. High level objects are, rather, configurations of low-level objects that gain and lose parts but maintain enough internal integrity to be tracked through change and reidentified across contexts.[4] This means that in order to know which patterns are functionally suited to play the role of 'objects' (the macroscopic things that we track visually and reidentify across contexts) at high levels of organization, it is not enough to know what the microscopic building blocks of matter are. One has to also know what kinds of patterns emerge when large numbers of those building blocks are put into interaction. Dynamics acquired a new importance in understanding how these high-level structures are stabilized out of low-level interactions, so formalizing the relationship between big things and little things, which was at first conceived as the *a priori* philosophical project of giving the logic of the

---

[1] "Physics and perception are like two people on opposite sides of a brook which slowly widens as they walk: at first is easy to jump across, but imperceptibly it grows more difficult, and at last a vast labor is required to get from one side to the other." (Russel 1992, p. 137), thanks to Dustin Olson for tracking down the quote for me.

[2] I use 'the manifest image of the world', 'the familiar world of everyday sense', and 'the world as we experience it' interchangeably. There are some distinctions we might want to make between these but they won't matter here. And I use the view of time sub specie aeternitatis and from a temporally transcendent perspective interchangeably.

[3] The properties of such things were known by their causal effects on macroscopic measuring instruments, and that raised issues about whether we had any direct grasp on the intrinsic properties of things. But structurally the macroscopic environment was thought to be a coarse-graining of the microscopic.

[4] A good example is provided by a wave moving across the surface of an ocean. The wave is a stable structure that can be identified and tracked as it moves towards shore. At any given time, it is wholly composed of water molecules, but there may be little or no overlap between the collection of water molecules of which it is composed at one time and that of which it is composed at another.

composition relation,[5] turned out to conceal a lot of interesting physics. The idea remained in place, however, that the manifest image is just a macroscopic coarse-graining of a universe described in microscopic detail by physics. And the difficulty was just trying to figure out which high-level configurations were going to be stable enough to act as targets for singular reference (i.e., to look suitably *thing-like* at the macroscopic level).

In prerelativistic days, *time* didn't seem to present any special difficulties. The time of Newton's physics was the dimension in which the History of the World unfolds, and it had the same dynamic character as our experience. But a whole new vision of time took shape with relativity that seemed to open a gap between the familiar time of everyday sense and time as it appears in physics. The new vision presented space and time together as a four-dimensional manifold of events, which came to be known as the Block Universe. In the Block Universe, there was no ontological difference between past, present, and future, and there was no process of coming into being. The universe simply *was*.

The difference between the familiar time of everyday sense and the Block Universe echoes an ancient debate between the Heraclitian and Parmenidean conceptions of the universe. The reaction against the new scientific image of time turned into a debate between two conflicting metaphysics: one that claims to have experience on its side, and one that claims to have physics on its side. Time as experienced is said to have four properties that the physical image of time lacks:

(v) Asymmetry: there are dynamical asymmetries in the behavior of macroscopic systems that make it easy to distinguish a film of everyday macroscopic processes run forward from their temporal reverse;
(vi) Flow: at any given moment, the world seems to be changing, or in flux;
(vii) Passage: when we look back over our histories, we see that what was once future is now present, and what was once present is now past;
(viii) Openness: at any given moment, there is one possible past and many possible futures.[6]

Giving explicit, non-metaphorical content to each of these properties is no easy matter. For historical reasons centering on the reduction of thermodynamics to the underlying microscopic theory, asymmetry became the focus of concerted attention in the foundations of physics. The topic remains one of the most active areas in foundational research. While there are many open questions, there has been a good deal of progress in finding a physical basis for the asymmetries that characterize the behavior of macroscopic systems. Passage and flow, by contrast, remain shrouded

[5] The name for this project was mereology, the theory of parts and wholes.

[6] Treat these as definitions that firm up terms that are often used loosely and interchangeably. 'Asymmetry' is often used to refer to any difference between past and future. I am using it to refer specifically to the dynamical asymmetries captured in the second law of thermodynamics. 'Passage' and 'flow' are often used interchangeably. As I use them, flow refers to how things seem at a given moment, whereas passage is a higher order comparison of how things seem at different moments. The point of that distinction emerges in connection with the question whether we perceive motion. No assumptions are made that this list is either exhaustive or exclusive.

in darkness. They are usually introduced with vague and poetic language. Openness rarely even warrants mention as something to which a sensible content can be assigned. Together, these aspects of temporal experience capture the Heraclitian vision of a universe in process, undergoing an absolute and irreversible process of *coming into Being*.

To many working in the foundations of physics, discussion of our experience of time is too imprecise and ill-defined to support real *research*. The most common reaction among those who are committed to physics as the source of ontological belief has been to dismiss the impression of passage, flow or openness as either nonsense or illusion: nonsense, because they are difficult to give non-metaphorical expression to; illusion, because there is nothing in physics that they would seem to describe.[7] But since experience is supposed to provide the evidence for our physical theories, physics can't ultimately avoid the need to connect itself to experience. The relationship between the flowing time of everyday sense and the static manifold of relativistic physics is one of the great, outstanding questions in our understanding of ourselves and our place in the universe.

In this paper, I outline a strategy for bridging the gap between the time of everyday experience and the time of physics which treats the Block Universe as a non-perspectival view of History and shows how to recover the everyday experience of time as a view of History through the eyes of the embedded, embodied participant in it. I also address questions about whether features of our temporal experience like passage and flow are properly thought of as illusory, the temptation to reify these features in the absolute fabric of the universe, and, finally, whether this strategy takes passage seriously.

6.1 Reconstructing Experience

We begin with some terminology. I use 'History' (capitalized) here to mean world-history, i.e., all of what happens everywhere from the beginning of time to the end. I use 'the view of History *sub specie aeternitatis*' to mean a representation of History whose content is invariant under transformations between temporal perspectives. The phrase 'sub specie aeternitatis' comes from Spinoza and it had a meaning for him that I don't want to take on board. The view sub specie aeternitatis, as it is intended here, is simply a representation of History that is not relativized to a temporal frame of reference. It captures only those intrinsic relations among events, independently of their relation to viewers, or anything else. The view sub specie aeternitatis is the temporal analogue of the view from nowhere. So conceived, the notion of the view sub specie aeternitatis is formally well-defined, though it is hard to find locutions that don't suggest perceptual metaphors that are somewhat

[7] Or worse, nonsensical. It is just as hard to characterize what these are supposed to mean in non-metaphorical terms, as it is to reconcile it with the relativistic image of time. There are some dissenters: Ellis (2008), John Norton (2010), and Smolin (2014).

inappropriate. So, for example, we speak of the '*view* sub specie aeternitatis' or the 'temporally transcendent *perspective*'. I'll continue to use these locutions, but I want to disavow the literal interpretation as point from which a space is *viewed*. The formal apparatus for talking about the relationship between frame-dependent and frame-independent representations is very well-developed, and the analogy with space is helpful to keep in mind.[8] When we talk about a perspectival representation of space, we give that content as representation of space that is implicitly relativized to a frame of reference defined by the observer's location and orientation in space. There is a logical transformation that takes us from a non-perspectival representation to a perspectival one and back.[9] In what follows, I give a similarly explicit characterization of 'the participant's perspective on History' and show how to obtain a *transformation* that takes a static image of a four-dimensional manifold into an evolving image of a universe in the process of Becoming.

Here is how the transformation is defined. We start with an account of how History looks from the perspective of a particular moment. Formally, the view of time from a particular moment is like the view of space from a particular location. Just as the view from here is a representation of a three dimensional manifold relativized to a reference frame defined by three points (one for each spatial dimension) in the space, the view from now is a representation of a linear order implicitly relativized to a point in it. The events of History are divided into three sets (past, present, and future) depending on their relationship to now. Different events are past relative to different moments in History, just as different points are nearby relative to different locations in space. The content of a representation of the world from a particular moment is like a snapshot of History taken from the here and now. The view of History over some interval—a day, a year, a life—is obtained by stringing together the snapshots from the moments that comprise the interval, in the order defined by their appearance in the interval. So, for instance, we get the view of a football game through the eyes of the wide receiver by stringing together the snapshots that represent his momentary perspectives from the beginning of the game to the end.

---

[8]There are many good discussions of reference frames in physics. For a nice philosophical discussion of the connection between invariance and objectivity, see Nozick (2001). The *locus classicus* of the philosophical discussion of the 'unembedded' or non-perspectival view of History see Nagel (1989) and Williams's (1976) remarks on the Absolute Conception of Reality. See also Ismael (2007) where the formal apparatus for talking about invariant content and the transition from embedded representation, whose content tends to be context-dependent, to forms of representation whose content is invariant under transformations between contexts.

[9]And in the spatial case, there is an object—the observer's body—that moves through the landscape as the frame changes. In the temporal case, whatever we mean by a temporal frame of reference, there is no object that moves through time as that frame changes. But even in the spatial case, the frame of reference is a relation between the contents of two kinds of representations: a visual representation in which space is represented in a manner that is relativized to a frame—either egocentric or allocentric, as the case may be. The viewer's map of her body and its location in space plays the role of the 'you are here' dot allowing her to coordinate visual information with spatial information (Klatzky 1998).

It is not trivial to get the content of these snapshots right. The natural thought is that the view of History from a particular moment is a combination of what the viewer is seeing at that moment and the contents of his memory, and that the perceptual part ('what the viewer is *seeing*') is a representation of the occurrent state of the environment. If that were right, perceptual representations themselves would represent instantaneous states of the world. Representations of motion and change, or temporal relations like before and after, or duration, would occur only in memory, where we construct representations that span longer intervals of history. And they would be inferred from comparisons of perceptually apprehended positions at different moments.

Many people nowadays reject this simple view of perception. Two primary arguments are given. One is the phenomenological evidence that motion is perceived directly, rather than inferred. What you see when watch a long pass tracing an arc across the sky is not a sequence of positions—the ball there, then the ball there, then the ball there—but a ball moving in a certain direction at a certain speed. Direction and speed belong not to points, but *intervals* of time.[10] The fact that you don't *infer* the direction and speed from a sequence of perceived positions, but *see* the direction and speed means that the *content* of even an *instantaneous* perceptual state spans a finite region of both space and time. The second argument is that perceptual illusions that have been well-documented in the lab bear out the idea that the brain is representing what happens over a temporal interval. Some experiments suggest that the interval extends (surprisingly) into the future.[11] If this is correct, then perception delivers not a sequence of static snapshots, but representations of movement and change. The intervals represented in perception, however, are very small. Estimates range from 25 to 240 msec. Much longer intervals are represented in memory and this is where, uncontroversially, most of the complex temporal content is contained.

Memory comes in numerous forms. Episodic memories take the form of recollected images, sights, sounds and smells. Autobiographical memory is devoted to the time-consuming work of constructing, interpreting, and condensing life experiences to produce a narrative account of a personal past.[12]

Perception and memory are both selective and reconstructive. There has been a lot of fascinating and somewhat surprising research in the last decade or two about the scope of the reconstructive nature of perception. It turns out that the brain does more than simply integrate information over a temporal interval. Instead, what you see is the prepared product of complex processing that involves filling in and projecting forward temporal inference. Between the moment the light hits your retina

[10] This is the proper way to understand William James' specious present. One has to be careful not to mistake the claim that the temporal content of perceptual representations spans a finite interval for a claim about how temporally wide the state itself is. This would be like saying that because a perceptual state represents an expanse of space, it must occupy that expanse. See Grush (2009).

[11] See Grush (2007) and Eagleman (2011).

[12] There is also semantic memory, muscle memory, and any number of others, which are less relevant for our purposes. See Sutton (2012) for a taste of the breadth of memory processing.

and your conscious awareness of the scene in front of you, there is a good deal of computation going on in your brain. This is a rich area of research that is turning up fascinating results.[13] With regard to memory, selection and reconstruction are less surprising, and there is a long history of research that confirms that there is a good deal of processing, and some confabulation, particularly in autobiographical memory.[14]

So far, we have been talking of our representations of the past. But we are forward-looking creatures and we represent the future as well as the past. Our representations of the future have a very different epistemic character from our representations of the past. There are two asymmetries in our relationship to past and future events. There is the epistemic asymmetry: we don't remember the future, so our expectations for the future are guesses at best, gleaned from information contained in perception and memory and eventually overridden by future experience. And there is the practical asymmetry:[15] since our beliefs about (some of) what happens in the future depend on what we decide, those beliefs about the future can't be settled until our decisions are settled. From the perspective of the decision-maker, making up her mind about what to do is also making up her mind about how the future will be.[16] When we represent History from the perspective of a particular moment, we see a fixed History, represented in a patchy way in memory, but beyond volitional control. When the decision-maker looks into the future, she sees a range of open possibilities whose resolution into fact hinges on decision.[17]

The asymmetries, as I have described them, are asymmetries in our epistemic and practical relations to the events being represented at different points in our lives. We can (and should) ask about the physical basis of these asymmetries, but for now we need to observe only that these practical and epistemic asymmetries are phenomenologically fundamental and structure our cognitive representations of the world. They form the practical and epistemic lenses through which we view the world. If we look at how our representations of time change as we run through the repeated cycle in which we preconceive our histories, plan, act, and feed the

---

[13] Dennett reported some of this work in Consciousness Explained. More recent work by Grush, Clark and Eagleman confirms and extends it.

[14] Gazzaniga (1998), and others. The word 'confabulation' suggests that memory is malfunctioning. That misses the point that autobiographical memory is not just a record of the past, but how we process information about the past for practical use. Telling the story of your past is a way of making up your mind about its significance. See also Schechtman (1996).

[15] We represent the future both in a passive epistemic mode (as when we are wondering, for example whether it will rain tomorrow), and in a deliberative mode (as when we are envisioning possible futures for ourselves and making decisions about how to act). These correspond to the two uses of "I think I am going to" in Anscombe's (1957) famous contrast between "I think I am going to be sick" and "I think I am going to take a walk".

[16] See Ismael (2011), also Velleman (1989), Joyce (2002), and Price (1992). In Ismael (2011) 'making up one's mind' is analyzed as a kind of mental performance. This imaginative picture is regimented formally in decision theory, in which the future is represented by a set of act-dependent possibilities, which are resolved into a singular outcome by the decision process itself.

[17] For the best, recent, book-length discussion of the nature of these asymmetries and their physical basis, see Albert (2000).

observed results of our actions into the next cycle of planning, we will find that the same events are represented from multiple perspectives: first in anticipation, later *in praesentia*, and finally in retrospect. If we look lengthwise over the course of a History, we see the changing perspective. And since memory is also keeping records of how the History looked through our eyes from one moment to the next, that structure is reproduced in every moment—like a set of nested snapshots of how History looked from the various temporal perspectives that make up our lives. The result is that the temporal content of an instantaneous cross-section of a normal human life is the accreted product of a more or less continuous cycle of reflective representation and re-representation in which perspectives are layered on top of one another. It is important to understand that we don't just represent the world. We represent our *own representations*,[18] capturing our epistemic and practical relations to what we represent, comparing our expectations with what actually happens, and opening up the space for complex attitudes like surprise, regret, disappointment, or relief, and making the change in our perspective, itself, something that is represented in thought, often as an object of poignant awareness. Throughout all of this, History itself is represented as the fixed object of representation. It is part of the content of our representation that the event anticipated is the *same* as the one experienced, and later remembered, and that what changes is our temporal perspective of the event; just as it is part of the content of our representation of a table as we walk around it that it is one and the same table that is seen now from this angle and now from that.[19] When people reflect on the passing of time, often they are calling attention to this change in perspective by looking back on events to which they earlier looked forward.

The upshot of all of this is that perception and memory working together produce an intricate structure of linked representations of the same moments in time, viewed from different perspectives over the course of a life. The human mind seems to be the only one whose representational states have this much explicit temporal content. Other creatures see movement and change, and other creatures seem to have map-like representations of *space*, but it is not clear whether there are other animals whose representational states have an explicitly articulated temporal dimension, i.e. an internal map-like dimension in which they store information about events when they are not happening. We may be the only ones, that is to say, who have a conception of History as it appears *sub specie aeternitatis*.[20] Just as we have a concept of space itself, independent of our relationship to it, we have a concept of History itself, independent of our relationship to it. History itself, or History viewed sub specie aeternitatis, is just what happens, a four-dimensional pattern of events. We can describe it back to front or front to back. It is not dynamic. It has no direction.

---

[18] The difference here is subtle but important. Think of the difference between a news report that simply describes the events of a battle, and one that reports on its reporting of the events.

[19] It needn't have been that way. We might have simply been aware of patterns of light and color. That wouldn't have been awareness of the world *as such*. There is little question that our spatial and temporal concepts have this much articulation.

[20] On the idea of an explicitly articulated temporal dimension, see Ismael (2007).

It is only when we look at how the world is represented in the representational states of a participant in it that we find the interpretations for flow, passage, and openness. The phenomenology of flow is a product of the way that the brain processes sensory information. The research on temporal illusions seems to confirm what we all know from experience, viz., that even the most rudimentary perceptual experience is an experience of change or motion. The sense of passage arises from the aforementioned poignant awareness of our changing perspective on history. Openness is a feature of the way that the future looks to the decision-making agent. From the perspective of such an agent, the decision process itself resolves a collection of open possibilities into singular fact.

The degree to which this strategy for reconciling human time and physical time is successful will depend on the degree to which it can faithfully recover the real, lived experience of the participant in History, and so these analyses of flow and passage and particularly openness have to be developed with some care. That is something I have tried to do in other places, but here I want to focus on the logic of the proposed relationship between the view sub specie aeternitatis and the view through the eyes of the participant. The claim is that in the view of History through the eyes of the embedded, embodied participant, events are ordered by their practical and epistemic relations to the viewer at different points in her life so that when they are strung together in a temporal sequence, they produce a changing image of a world with a fixed past and open future, in the process of coming into Being. Passage, flow, and openness arise as artifacts of changes in perspective, relative to the fixed backdrop of History. In the view sub specie aeternitatis, by contrast, those same events are represented in a way that is invariant under transformations between temporal perspectives. This doesn't mean that the practical and epistemic asymmetries disappear, but their relational character is now made explicit in precisely the same way that when we move from a perspectival representation of space to a map-like representation, relations like 'nearby' are explicitly relativized to spatial perspectives. And we can transform between the view sub specie aeternitatis and the view through the eyes of the participant in History in the way we can transform between egocentric and map-like representations of space.[21]

6.2 Closing the Circle: From Thinking inside Time to Thinking outside Time and Back[22]

There is a lesson in all of this that bears on my opening remarks pertaining to the relationship between the manifest image and the scientific image of the world, which is to highlight the broadly logical suggestion that the reason that time has

[21] And from a relativistic perspective, of course, space and time are united in the Block Universe and perspective is conceived as the here-now of located experience.

[22] The phrases 'thinking in time' and 'thinking out of time' are introduced by Smolin (2013).

seemed so hard to accommodate is that there is a crucial component in the relationship between the manifest and scientific image that has been left out.

Physicists have focused a good deal on transforming a fine-grained macroscopic model into a coarse-grained image, but the task of transforming a view of time sub specie aeternitatis into a view through the eyes of the embedded, embodied *participant* in history (in ways that explicitly recognize how events are ordered in her experience and by her practical and epistemic relations to them) has remained out of focus. I tend to think of these in somewhat picturesque terms as two separate dimensions that have to be bridged in relating physics to phenomenology.

The reason that physics has done a decent job accommodating asymmetry, but not such a good job with flow, passage, and openness, I would suggest, is that asymmetry is an artifact of the shift from a microscopic to a macroscopic perspective, whereas flow, passage, and openness arise in the transformations wrought in that horizontal dimension. Adding the horizontal dimension allows us to close the circle, bringing experience and ontology back together as part of a single, unified vision of the universe in which experience furnishes information about ontology and ontology includes experience.[23]

Those familiar with Hartle's paper "The Physics of Now" will recognize from his discussion the seeds of this strategy for reconciling the relativistic image of time with our temporal experience. In that paper, he showed how to find an interpretation of the distinction between past, present and future in the representational states of a system whose practical and epistemic perspective mirrors our own (i.e., in a robot with a memory and sensors that moves around the world gathering information and using it to guide behavior). One of the reasons that Hartle's IGUS made an important impact in the physics literature on time, is that it is an effective tool for bringing issues about experience back into the fold of physics without getting caught up in the philosophical tangles associated with mental phenomena. The IGUS provides something purely objective that can serve as a kind of bridge between the dialectical worlds of figures as different as Einstein and Bergson.[24] When Bergson talks about human experience, he will want to talk about something identified by the role it plays in human life that Einstein will want to dismiss as outside the purview of physics. But if we can identify representational states inside an information-gathering and -utilizing device like a robot, which at least have the same functional role as the progression of states that constitute our conscious mental lives, then we can *locate* something that even Einstein will have to recognize falls within the purview of physics, and we will have found some common ground. Now we have a two-part story. The first part of that story is recognizable as physics. It describes the emergence of the thermodynamic gradient and the dynamical asymmetries that characterize the observed world. The second part of that story is less recognizably *physics*, though it is of a piece with the physical story. It is the story of how the thermodynamic gradient paved the way for the emergence of information-gathering and -utilizing systems and how the world is represented in the internal states of

[23] Closing the circle, in Shimony (1993).

[24] See Canales (Chap. 4, this volume).

those systems. Asymmetry arises at the first stage. Passage, flow, and openness arise at the second stage.

Just as there is no need (or warrant) for reifying at the fundamental level, those features of the observed world that are generated at the first stage, there is no need (or warrant) for reifying, in the absolute structure of time, features of experience that are generated in the second stage. At the fundamental level, we have the static four-dimensional manifold with only those temporal asymmetries that are dictated by our microlaws. At the macroscopic level we have the thermodynamic gradient. At the level of human psychology, we have the flowing, directed time of everyday sense.

6.3 Relative Versus Absolute Becoming

This way of reconciling the Parmenidian and Herclitian visions of time also provides a formal resolution to the logical puzzle presented by McTaggart's argument, i.e., the puzzle of how to integrate the A-series with the B-series.[25] The B-series is the set of moments of history ordered by relations of temporal precedence. The A-series is the set of moments divided into past, present and future, hence ordered by their relation to the present moment. McTaggart argued that the two series' could not be integrated, and hence that the very same moments that had fixed locations in the B-series could not consistently be regarded as ordered by their relations to the present moment. The conclusion of his argument was that either (i) A-series properties are implicitly relativized to B-series locations, in which case the B-series is (really) all there is, or (ii) we get a contradiction. The strategy I have proposed comes down firmly on the side of 'The B-series is all there is, in the absolute, non-relational structure of time'. A-series properties are included implicitly as B-series properties relativized to a complex, evolving perspective.

Although the underlying logic is complex, the view just formalizes things that, I would argue, we all know. We know that we can willfully affect things that lie in our future, but not things that lie in our past. We know that we can remember things that lie in our past, but not our future. And we know that the practical and epistemic asymmetries that characterize the view of history from a particular moment are perspectival, in the sense that they are different at different moments in our lives. An event that is anticipated at one time is remembered at another. An event that is open (i.e., within practical reach) at one time is fixed (beyond practical reach) at another. These differences are not intrinsic to the events that constitute History, but differences in the relations that we bear to those events at (or from) particular moments in our lives. The way that we integrate information about History over time, moreover, makes it clear that we know the difference between structure that represents the intrinsic ordering among events, and structure that is relative to a spatial or temporal perspective. We know that the division into past and future is relative to a moment.

[25] McTaggart (1908).

We know that the same event is future at one moment, present at another, and past at others. When we integrate information about events across multiple perspectives, we identify yesterday's tomorrow with tomorrow's yesterday and understand how our relationship to the day has changed in the interim. When we update our beliefs, we make it clear that we understand that what could have been avoided yesterday is now beyond avoiding. In short, everything about the way that we manage our beliefs about the world makes it clear that we understand that the division between past and future, together with all of the practical and epistemic asymmetries that that division imposes on our relations to events, is perspectival.

When History is represented sub specie aeternitatis we are forced to relativize the perspectival structure to make it invariant under transformations between momentary perspectives, and so we make explicit what we all know in the separation of perspectival structure from structure that is intrinsic to time. The practical reason that we go in for representing time in a manner that is invariant under transformations between temporal perspectives is that doing so supports planning. In order to form a coordinated, temporally extended plan of action, one has to map out the parts of the action and keep track of one's progress.[26] This form of representation, however, is not telling us anything that we don't already (at least implicitly) know about the metaphysical status of features like past-ness, presentness, fixity and openness.

It has been customary for those that accept Parmenidean metaphysics of time to reject passage, flow, and openness as illusory, often citing Einstein's famous remark about the distinction between past, present, and future being a stubbornly persistent illusion.[27] The thought seems to be that if passage, flow and openness are mere artifacts of perspective, they aren't 'real' or objective.[28] I don't see that this talk of illusion can withstand scrutiny. Perspectival structure is recovered in the view of time sub specie aeternitatis as explicitly relational, on precise analogy with the spatial case. No structure is lost. Passage, flow and openness remain as real as the difference between nearby and far away, the distinction between up and down. The lesson is not that Parmenideans win and Heraclitians lose. It is that there is no genuine conflict. The view sub specie aeternitatis includes the evolving view of time pre-

[26] See Bratman (1987) on time, planning and agency.

[27] In other moods, Einstein took it quite seriously. Carnap reports that: "Once Einstein said that the problem of Now worried him seriously. He explained that the experience of the Now means something special for man, something essentially different from the past and the future, but that this important difference does not and cannot occur within physics. That this experience cannot be grasped by science seemed to him a matter for painful but inevitable resignation. I remarked that all that occurs objectively can be described in science; on the one hand the temporal sequence of events is described in physics; and on the other hand, the peculiarities of man's experiences with respect to time, including his different attitude towards past, present and future, can be described and (in principle) explained in psychology. But Einstein thought that these scientific descriptions cannot possibly satisfy our human needs; that there is something essential about the Now which is just outside the realm of science" (1963, p.37).

[28] Even with the good guys, people like Craig Callender (2010); Sean Carroll (2010), who agrees, in outline, about where an explanation of the experience of passage should come from, the vocabulary of illusion remained firmly in place.

sented in the experience of the participant in History as a view through the eyes of the participant in History, with the practical and epistemic lenses imposed by the physics of the environment and the terms of her embodiment. The view through the eyes of the participant includes the view sub specie aeternitatis as the invariant relation among the parts of time.[29]

On this view, there is no relevant logical or ontological difference between the status of the Heraclitian properties of flow, passage, and openness and perspectival spatial properties like the far away-ness of Neptune or the motion of the houses lining a street when viewed through the window of a moving train. When we move to a representation whose invariance class includes a parameter, P, structures that were absolute before the move get explicitly relativized to P-values. No structure gets demoted from 'real' to 'unreal'. We simply have a representation of the structure that separates the absolute from the relational and makes the relational character of P-relational structures explicit. To think that accepting the Block Universe as an accurate representation of time as it appears sub specie aeternitatis means rejecting passage, or flow, or openness, as illusory is like thinking that accepting a map as a non-perspectival representation of space means that you are under an illusion that anything is nearby.[30] As we develop an increasingly absolute conception of the world, more and more of the structure at the forefront of our experience of the world is revealed to be perspectival. It's difficult to say how 'perspectival' came to be associated with 'unreal',[31] but that association has been one of the most insidious and confusing aspects of the physical discussion of time. I have emphasized that perspectival structure of the kind that is being discussed here is perfectly real; it is just implicitly relativized to distinctions introduced by the agent's perspective.

---

[29] Sometimes people speak as though the defenders of passage are just making the mistake that if they see a world line written down on a piece of paper, it doesn't look like it is changing, so they reject the view that change is just having different properties at different times. Of course, that is a mistake. We can represent change by stringing representations of moments together in a temporal sequence, but we can also represent it by arranging representations of moments lengthwise along a page with the temporal parameter represented by the horizontal dimension along the page, or by writing down a mathematical function that represents evolution with respect to time. But to think that is the mistake that is always in play underestimates the problem. The problem is that we need to get flow and passage and openness into the content of experience without reifying them in the absolute fabric of the world.

[30] The logic of the relationship is a little complicated, because time is both what is being represented in the content and defining the frame from which it is represented, so we get the impression of the events of History being ordered and reordered by their relations to an object—the now-moving through time. For more on this see Chap. 10, Ismael (2007). The technical resolution is that the now is not an object, but the fixed point in a series of frame-dependent representations of time that has different values for different elements in the series.

[31] It may be an artifact of the tangled history of coordinate systems in physics. 'Perspectival' came to be associated with 'coordinate-dependent' which is used to identify aspects of mathematical representations of space-time that have no physical significance. There are many excellent accounts of that history. See especially Friedman (1983). Or perhaps it was because perspectival means implicitly relational, and hence neither absolute nor fundamental. But the 'real' is surely not coextensive with either the absolute or the fundamental.

There *is* an illusion if we treat perspectival structures as absolute, that is, if we reify structures that belong properly to the perspective of the participant in History in the absolute fabric of space and time. So, for example, if we treated the division into past, present, and future, and the practical and epistemic asymmetries that go with that division, as intrinsic features of events, we would be subject to an illusion. Who makes this mistake? If the metaphysics of common sense are culled from everyday practices of integrating temporal information over time, we can't convict common sense of this mistake. It is, however, what happens when common sense begins to philosophize, or when we take the little pictures that people carry around in their heads as metaphysical committments. It is likely true that the man on the street carries around a picture in his head of a universe unfolding as he experiences it. But it is also true that the way he integrates temporal information across perspectives shows that he is *not* subject to that illusion. He uses calendars and time-lines unproblematically, and probably doesn't spend much time worrying about how to fit the two pictures together.

McTaggart's argument was intended to show that common sense has an incoherent metaphysics of time. I think that what it actually shows is that common sense doesn't have an articulate metaphysics of time, and so can be easily drawn into contradictions. When the man on the Clapham omnibus is forced into metaphysical commitments by an insistent questioner, or when the philosopher tries to form an explicit response to McTaggart's argument, he gets tied up in knots. But the *pre-philosophical* phase of temporal thought, it seems to me, is fine. McTaggart's argument initiates a deeply confused philosophical phase, because the logical structure of beliefs about time is quite complex. This progression from unreflective common sense, through philosophical perplexity, to an articulate metaphysics is characteristic of the sorts of problems that arise when common sense is put under philosophical pressure.[32] It is a lovely illustration of the dual role of philosophy, leading first *into,* and then (one hopes) *out of,* confusion.

6.4 "Taking Passage Seriously"

The relational view is often said to "not take passage seriously". This charge is made, for example, in a recent paper by Pooley. Here is the abstract::

> Is the objective passage of time compatible with relativistic physics? There are two easy routes to an affirmative answer: (1) provide a deflationary analysis of passage compatible with the Block Universe or (2) argue that a privileged global present is compatible with relativity. (1) Does not take passage seriously. (2) Does not take relativity seriously.[33]

---

[32] I recognize, of course, that the line between carrying around a mental picture and elevating it to the status of a metaphysical view is a very fine one, and whether there really are any philosophical innocents is a real question. Whether my pre-philosophical man—my man on the Clapham omnibus—is a mythic figure or a real one doesn't matter for our purposes here, but I think that philosophers are overly inclined to think that everyone is a metaphysician. I think that many of the people I know best never asked the question "What is time?" in a form that demands a metaphysical answer. And I think the pre-philosophical phase is a fine one to remain in.

[33] Pooley (2013).

If taking passage seriously is a matter of insisting that passage is a real feature of our temporal experience, and demanding that physics be able to account for it, then the view I have argued for takes passage very seriously indeed, but it does not satisfy Pooley's definition of taking passage seriously. Part of the reason that one might deny that the view takes passage seriously comes from the idea that on a relativized view passage turns out to be 'illusory'. I have said why I think it is mistaken. Certainly it turns out to be perspectival, but if there is one lesson to be emphasized, it is that perspectival structure is not always 'illusory'. It is an important part of the relationalist view that we do not have to reify the relational in order to regard it as real. Nor do we need to reify it to regard it as worthy of a distinguished role in human life. It makes perfect sense that our cognitive and practical lives should be organized around distinctions that have a merely relational significance. We care more about what is nearby than what is far away, not because what is nearby is intrinsically ontologically special, but because it is *nearby*. We care more about our own children than other people's children, not (or not just) because they are more intrinsically special than other people's children, but because they are *ours*. Indeed, I think that everything that we care about is at the interface between what (sub specie aeternitatis) is the case and *how* what (sub specie aeternitatis) is the case relates to the here and now, to ourselves, and to our place in History. To be human is both to have an eye on eternity and feet in the here and now. By 'taking passage seriously', Pooley means not simply reconstructing passage as a feature in the experience of embedded agents, but underwriting a view according to which the universe itself is undergoing a process of coming into Being. I'll designate taking passage seriously in this sense, taking passage SERIOUSLY. Relationalists generally register puzzlement about what the idea is. For the relationalist, the Block Universe simply formalizes the recognition that the distinction between past, present and future (and all of the asymmetries that attach to that distinction), is relative to a moment in time. There is a good deal of talking past one another in the literature, with relationalists reconstructing all forms of passage in relational terms and opponents denying that this is what they mean.

Although the Block Universe is by far the dominant picture that one finds in physics textbooks, views that purport to incorporate non-relational forms of passage have recently made it into literature in the foundations of physics, conspicuously in the view of Smolin. The issue for Smolin focuses on the status of a global present, and to see what he has in mind, we need to look at the transition from Newtonian to Relativistic physics. In Newtonian physics, the notion of the state of the world at a time is well-defined and absolute. We can speak in a non-perspectival way of the state of the world at one time and represent the History of the universe as a sequence of states, one followed by the next. In this picture, time is treated as an external parameter in which the history of the world unfolds. The division between space and time is objective in the sense that it is not simply relativized to a point in a psychological history, but the history of the universe itself is described as a sequence of historical stages. In Special Relativity (STR), there is no longer the separation of space and time that allowed us to treat time as an external parameter. Spatiotemporal intervals are absolute; but spatial and temporal intervals are not. The distinction between temporal and spatial structure is not drawn globally, but locally, by the light-cone structure at each point in space-time. Temporal order and duration are defined along a time-like curve. For any inertial trajectory there is a method for extending these local notions off the curve to a global notion of time, but if we compare the global notions, we will find that the global notions associated with different inertial trajectories disagree on which events are happening at a given moment. So judgments about which events are happening at a given moment are perspectival in STR but not in Newtonian physics. This is put by saying that there is no absolute notion of simultaneity in Minkowski space-time, or that there is no globally defined notion of the present state of the world. The story is a little more complicated in the General Theory of Relativity. There are no global inertial frames in a generic general-relativistic space-time. But in a neighborhood of any space-time point there is a continuous time-like curve in whose neighborhood one can define what is called a normal frame, which specifies a privileged family of instantaneous three-dimensional spaces and says what events in different instantaneous spaces occur at the same enduring place. The family, however, may not be defined globally. So while there is a well-defined frame in the neighborhood of every point in which we can talk about the state of the world, the state of the world at a time is still perspectival. The upshot is that, while the division between past, present and future was perspectival in Newtonian physics, in the sense that it was relativized to a moment in time, it is perspectival in a new and stronger way in relativistic theories. In relativistic theories, the distinction between past, present, and future is relativized to a point along a time-like curve. Why does this matter? As long as there was a globally defined present, it was possible to invest the distinction between past, present and future with ontological significance, i.e. to think that it marks a distinction between what is real and what is not. In STR the manifold does not have enough invariant structure to support the ontological weight of that distinction.

The reconstruction of temporal experience that I have proposed separates the question of temporal passage—understood as something of which we are immediately aware in experience—from the question of whether there is a globally defined present. It treats this last as a question for physics, to be settled by considerations of a kind that are far removed from everyday experience. On this view, we have an internal time, defined for the psychological history of embedded observers by the flowing, passing character of everyday experience. Communication among observers (and the creation of time-keeping technologies like clocks and watches) will stabilize an intersubjective notion of 'what time it is' well enough for practical purposes, but leaves questions about the absolute structure of space and time to physics. It takes passage seriously in that it insists that, in order for our physics to provide an intelligible picture of ourselves and our place in nature, we need it to support real, lived everyday experience. But it denies that physics has to confirm folk ontology. By showing how to take passage seriously without taking passage SERIOUSLY, it shifts the burden of argument. Physicists like Smolin do take up that burden, marshaling considerations drawn from physics in support of the existence of a global present. But at this stage, the question is straightforwardly a question of physics. It has nothing directly to do with temporal experience.[34]

[34] Nor does it seem to have anything to do with the Block Universe. It is not, for example, that a Block Universe is incompatible with the existence of a global present. The Block Universe is just

6.5 Concluding

I have defended the Block Universe as an image of History as it appears sub specie aeternitatis and described a strategy for recovering the everyday experience of time as a view of time through the eyes of the embedded, embodied participant in History. I addressed the common misconception that perspectival structure is illusory and denied that a view that treats passage as perspectival fails to take it seriously. There remain many open questions about time in physics. There are questions about, for example, whether there is a globally privileged present[35] or a fundamental global form of 'becoming', or whether space-time is itself emergent from a non-spatiotemporal structure.[36] But the methods for answering these questions have nothing directly to do with reflecting on the character of our temporal experience. They are questions about whether these elements of structure are implicated in the movements of objects, i.e., whether there are law-governed differences in behavior explained by differential relations to those structures.

Works Cited

Albert, D.Z. 2000. *Time and chance*. New York: Harvard University Press.

Andersen, H.K., and R. Grush. 2009. A brief history of time-consciousness: Historical precursors to James and Husserl. *Journal of the History of Philosophy* 47(2): 277–307.

Anscombe, G.E.M. 1957. *Intention*. Cambridge, MA: Harvard University Press.

Bratman, M. 1987. *Intention, plans, and practical reason*. Cambridge, MA: Harvard University Press.

Callender, C. 2010. Is time an illusion? *Scientific American* 302(6): 58–65.

Callender, C., and C. Wüthrich. 2014, 1 March. What becomes of a causal set. (Preprint draft). Retrieved from http://philosophyfaculty.ucsd.edu/faculty/wuthrich/pub/CallenderWuthrich 2014Causets_AprilWe b.pdf

Carnap, R. 1963. *The philosophy of Rudolf Carnap*, ed. P.A. Schilpp. New York: Cambridge University Press.

Carroll, S. 2010. *From eternity to here: The quest for the ultimate theory of time*. New York: Dutton/Penguin Group.

Eagleman, D.M. 2011. *Incognito*. New York: Pantheon Books.

Eagleman, D.M., and J. Downar. 2011. *Cognitive neuroscience: A principles-based approach*. New York: Oxford University Press.

Ellis, G.F. 2008. *On the flow of time* (No. arXiv: 0812.0240 [gr-qc]).

Friedman, M. 1983. Foundations of space-time theories. Princeton: Princeton University Press.

a generic framework that can support any absolute structure that can be defined on a four-dimensional manifold. Smolin associates the Block Universe with the reality of the future and is concerned to deny that the future is real. But again, here, one feels that there is some talking past one another. Surely events that are future now will be present later, and so questions of what is real have to be relativized along with the distinction between past and future, and the Block Universe is entirely compatible with that.

[35] See Maudlin (2007).

[36] Huggett and Wüthrich (2013).

Gazzaniga, M.S. 1998. *The mind's past*. Berkeley: University of California Press.
Grush, R. 2007. Time and experience. In *The philosophy of time*, ed. T. Müller. Klosterman: Frankfurt.
Grush, R. 2009. The temporal content of perceptual experience. In *The Routledge companion to philosophy of psychology*, ed. J. Symons and Calvo Paco. New York: Routledge.
Huggett, N., and C. Wuthrich. 2013. The emergence of spacetime in quantum theories of gravity. *Studies in History and Philosophy of Modern Physics* 44(3): 273–275.
Hunemann, P. forthcoming. Decision and the open future. In *The future of the philosophy of time,* ed. Adrian Bardon. New York: Routledge.
Ismael, J. 2007. *The situated self*. New York: Oxford University Press.
Ismael, J. 2010. Temporal experience. In *The oxford handbook of philosophy of time*, ed. C. Callendar. New York: Oxford University Press.
Ismael, J. 2011. Decision and the open future. In The future of the philosophy of time, ed. Bardon Adrian. New York: Routledge.
Ismael, J. forthcoming. Passage, flow and the logic of temporal perspectives. In *Cosmological and psychological time, The Nature of Time, The Time of Nature*, eds. C. Bouton and P. Hunemann. Chicago: University of Chicago Press.
Joyce, J.M. 2002. Levi on causal decision theory and the possibility of predicting one's own actions. *Philosophical Studies* 110(1): 69–102.
Klatzky, R.L. 1998. Allocentric and egocentric spatial representations: Definitions, distinctions, and interconnections. *Spatial cognition, January,* 1–17. Berlin: Springer.
Maudlin, T. 2007. *The metaphysics within physics*. New York: Oxford University Press.
McDaniel, K. 2013. John M.E. McTaggart, In *The Stanford encyclopedia of philosophy* (Winter 2013 Edition), ed. E.N. Zalta. http://plato.stanford.edu/archives/win2013/entries/mctaggart/
McTaggart, J.E. 1908. The unreality of time. *Mind* 68(17): 457–474.
Nagel, T. 1989. *The view from nowhere*. New York: Oxford University Press.
Norton, J.D. 2010. Time really passes. *Humana Mente: Journal of Philosophical Studies* 13: 23–24.
Nozick, R. 2001. *Invariances: The structure of the objective world*. Cambridge, MA: Harvard University Press.
Perry, J., and S. Blackburn. 1986. Thought without representation. *Proceedings of the Aristotelian society, supplementary volumes,* 137–166.
Pooley, O. 2013. Relativity, the open future, and the passage of time. *Proceedings of the Aristotelian Society* 113(3): 321–363.
Price, H. 1992. The direction of causation: Ramsey's ultimate contingency. In *PSA: Proceedings of the Biennial Meeting of the Philosophy of Science Association,* 253–267. Philosophy of Science Association.
Rea, M. (ed.). 1997. *Material constitution: A reader*. New York: Rowman & Littlefield Publishers.
Russell, B. 1927. *The analysis of matter*. New York: Harcourt, Brace & Company.
Schechtman, M. 1996. *The constitution of selves*. Ithaca: Cornell University Press.
Shimony, A. 1993. *The search for a naturalistic world view*, vol. 1. New York: Cambridge University Press.
Smolin, L. 2014. *Time reborn*. New York: Houghton Mifflin Harcourt Publishing Company.
Sutton, J. 2012. Memory. In *The Stanford encyclopedia of philosophy* (Winter 2012 Edition), ed. E.N. Zalta. http://plato.stanford.edu/archives/win2012/entries/memory/
Velleman, J.D. 1989. Epistemic freedom. *Pacific Philosophical Quarterly* 70(1): 73–97.
Williams, B. 1976. *Problems of the self: Philosophical papers 1956–1972*. New York: Cambridge University Press.

Chapter 7
Relation, Action and the Continuity of Transition

Tamar Levanon

> *In what sense can the parts = the whole? Question to be resolved dynamically, functionally, dynamically.*
>
> *(William James, "The Miller-Bode Objections")*
>
> *There is no nature apart from transition.*
>
> *(Alfred North Whitehead, Modes of Thought)*

Abstract This paper addresses a question that arises while attempting to conceptualize temporal experience, that is, the experience of continuity and transition. The question is how to handle the tension between the unity of the experienced flow on the one hand and the internal variation that it involves on the other. These two features which naturally integrate within the scope of our *experience* of temporality are incompatible in the framework of a more systematic analysis. It is this gap between experience and analysis which underlies the dispute between Bertrand Russell, William James and Alfred North Whitehead over a theory of continuity. More specifically, the dispute is over the role that our crude feeling of continuous transition should play within the conceptualization of a more general theory of continuity. While Russell thought that the intuitive appeal to our feeling in the formulation of a theory of continuity must be rejected, James' and Whitehead's accounts of temporal continuity are not only consistent with our feeling but actually *arise* from it. A comparison of their different perspectives reveals the role such concepts as succession, duration, simultaneity, unity and multiplicity in the formation of a theory of continuity.

Keywords Continuity • Temporal-consciousness • Specious present • William James • Bertrand Russell • Alfred North Whitehead

T. Levanon (✉)
The Department of Philosophy, Bar-Ilan University, Ramat Gan 5290002, Israel
e-mail: levanon.tamar@gmail.com

Y. Dolev, M. Roubach (eds.), *Cosmological and Psychological Time*, Boston Studies in the Philosophy and History of Science 285,
DOI 10.1007/978-3-319-22590-6_7

7.1 Introduction: Whitehead, James, and the Intuitive Definition of Continuity

In "A World of Pure Experience", William James describes temporal experience in the following manner: "What I do feel simply when a later moment of my experience succeeds an earlier one is that though they are two moments, *the transition from the one to the other is continuous*. Continuity here is a definite sort of experience; just as definite as is the discontinuity-experience which I find it impossible to avoid when I seek to make the transition from an experience of my own to one of yours" (1904, 536; emphasis added). This intuitive definition of continuity describes the *smooth transition* we experience when successive moments join together to create a unified stream. It involves a multiplicity of moments "without breach, crack or division" as James defines it elsewhere (1890, 237).

However, in attempting to provide a more systematic analysis of the continuity that characterizes temporal experience we encounter the problem of how to resolve the tension between the unity of the flow on the one hand and its internal variation associated with the idea of succession on the other. The problem stems from the fact that neither the idea of a multiplicity of elements ordered in succession nor that of a whole temporal interval clarifies the process of unfolding moments that creates the complex yet unified whole, a process which we refer to as "continuous transition". The attempts to determine the priority of unity over multiplicity or vice versa simply fail to capture the experience of continuity *as* transition or flow. For example, if we begin our investigation of continuity with succession, that is, with moments ordered before and after one another, then the continuous transition we experience eludes us as something that happens between successive moments (James 1909, 236). On the other hand, if we begin our investigation of continuity with a unified temporal stretch, then the distinction between its earlier and later phases becomes obscure. The duality that characterizes temporal experience is embedded in James' formulation of the difficulty as "the general conceptualist difficulty of any one thing being the same with many things, *either at once or in succession*" (1909, 281; emphasis added). James provides an additional formulation of the problem in "The Miller-Bode Objections" (1988a, 116), in which he asks the following:

> Whence the *co* if we start pluralistically?
> " " *ex* " " " mon " " ?

The following analysis attempts to tackle this problem and in particular examines the role that our intuitive description of temporal experience plays in various attempts to theorize the idea of continuity and to reconcile its different aspects. To this end, the different solutions proposed by James and two of his contemporaries—Alfred North Whitehead and Bertrand Russell—will be considered. They all agree that a theory of continuity must be consistent with our experience and capable of expressing it; however, Russell's view is contrary to those of James and Whitehead. Following his sweeping denial of change "in the metaphysical sense" (1903, 478), Russell tries to show that the intuitive appeal to the feeling of transition when formulating a theory of continuity must be rejected. Russell's rejection of the intuitive description of continuity involves two aspects: structural and dynamic. First, Russell denies transition in the sense of a unifying element that somehow connects different things and second, he denies transition in the sense of an intrinsic property, i.e. a dynamic element inherent in things. I discuss these two aspects of Russell's denial of transition in the second part of the paper. This provides the background for clarifying (in Sect. 7.3 and 7.4) James' and Whitehead's alternative accounts of continuity *as* transition.

James' and Whitehead's reasoning takes the opposite direction from that of Russell since both attempt to construct an account of temporal continuity that is not only consistent with our experience but actually *arises* from it. For both, it is experience itself that needs to be investigated and it is through the analysis of its elements that we gain our understanding of reality. Despite their different styles, it is this common mindset that characterizes both systems. James' emphasis on the role of experience is better known; he views the flux of sensible experience itself as containing "a rationality that has been overlooked, so that the real remedy would consist in harking back to it more intelligently, and not in advancing in the opposite direction" (1909, 73). In *Some Problems of Philosophy* he writes: "If the aim of philosophy were the taking full possession of all reality by the mind, then nothing short of the whole of immediate perceptual experience could be the subject-matter of philosophy, for only in such experience is reality intimately and concretely found" (1911, 53). Whitehead's description is less well known and is worth briefly dwelling upon. In *Process and Reality*, Whitehead argues that the meaning of "togetherness" is always drawn from "experiential togetherness", that is, the unique "togetherness of the component elements in individual experience" (1929, 189). According to Whitehead, any attempt to deny the priority of experiential togetherness:

> [...] leads to the disjunction of the components of subjective experience from the community of the external world. This disjunction creates the insurmountable difficulty for epistemology. For intuitive judgment is concerned with togetherness in experience, and there is no bridge between togetherness in experience, and togetherness of the non-experiential sort. (1929, 190)

According to Whitehead's "speculative philosophy"—which is the attempt to frame a "coherent, logical, necessary system of general ideas in terms of which every element of our experience can be interpreted" (1929, 3)—experience is not something that needs to be illuminated from an external, objective point of view. Rather, it is something that directs our inquiry. Thus, Whitehead's metaphysics is an adaptable, non-absolute picture, quite different from the traditional way in which we usually think of metaphysics. Whitehead's position is further clarified in light of his denial of what he calls "the bifurcation of nature." By using this phrase, Whitehead posits the separation of human experience from external things, or in other words the splitting of reality into two distinct systems. As he puts it, "what I am essentially protesting against is the bifurcation of nature into two systems of reality, which, in so far as they are real, are real in different senses" (1920, 30).

Indeed, Whitehead understands nature as a closed system "whose mutual relations do not require the expression of the fact that they are thought about"; thus, "in a sense nature is independent of thought" and "we can think about nature without thinking about thought" (1920, 3). However, these statements do not deny the fact that nature is also that which is perceived and thus "for natural philosophy everything perceived is in nature. We may not pick and choose" (1920, 29). More precisely, what we perceive is something real which is not reducible to a mere thought; yet at the same time our perception of nature is also a part of nature. The philosophy of nature "should never ask, what is in mind and what is in nature" (1920, 30) since our experiences of the apparent world are also nature itself (1922, 5).

7.1.1 The Specious Present and the Dual Appearance of the Tension between Unity and Multiplicity

Another idea which adds to the complexity of James' "general conceptual difficulty" is that of the specious present, i.e., the idea of a temporal stretch which is experienced as present. Russell, James and Whitehead all incorporate this idea within their analysis of temporal experience. However, the idea of the specious present, which is justified primarily on phenomenological grounds, is associated with James and Whitehead due to the role it plays in their more general discussions of continuity. For both, an extensive minimal unit of duration is the initial datum that is perceived in its entirety as an undivided temporal stretch, thereby enabling the experience of transition. Like a note of music, this ultimate datum requires "its whole period in which to manifest itself" (Whitehead 1926, 38). The best known formulation of this idea is provided by James in his *Principles*. In contrast to the "strict present", an ideal abstraction whose existence "can never be a fact of our immediate experience" (James 1890, 609), the immediate givenness of the specious present is emphatically emphasized by James in Chap. 15. In *A Pluralistic Universe*, similar claims regarding the direct experience of a minimal unit of temporal extension—for example, "the tiniest feeling that we can possibly have comes with an earlier and a later part and with a sense of their continuous procession" (1909, 282)—are further backed up by the more metaphysical argument against the infinite divisibility of time (1909, 228–231).[1] Whitehead willingly embraces James' conclusions which are "required by the consideration of Zeno's arguments, in connection with the presumption that an actual entity is an act of experience" (1929, 68).[2]

However, this description of temporal continuity, the essence of which is based on the succession of units of duration, actually reveals *both* sides of the problem, each of which now appears at a different level of analysis. For once we admit that

[1] Although James retained the idea of minimal units of temporal extension, he did not use the term 'specious present' in *A Pluralistic Universe*.

[2] Using 'actual entities', Whitehead specifies the metaphysical building block of his system. They are further defined as "drops of experience, complex and interdependent" (1929, 18).

the units of duration are ordered in succession, the experience of transition that we are trying to define disappears between those units once again. Moreover, the difficulty now appears in a new form, this time in relation to each unit, since *as* a temporal stretch it seems to contain both earlier and later phases. As such, its analysis should be in terms of succession. Only now these successive phases are all thought of as contained in "the present". This problem, which concerns the internal structure of each experiential unit, is sometimes formulated as the question of how successive phases are experienced together, and at other times as the question of how simultaneous phases can give rise to succession and temporality. For example, Max Blake writes in response to James' and Whitehead's common appeal to the idea of the specious present: "I fail to understand how a whole, every part of which is simultaneous with every other part, can be properly characterized as 'duration', or can be said to involve 'a definite lapse of time'" (1926, 649). A different formulation of this question was recently provided by Sean Kelly who complains that "the Specious Present is not about simultaneity; it is about temporal extent. The doctrine is committed not to the claim that we experience temporally distinct events as simultaneous, but rather to the claim that we experience temporal extension itself. What could count as evidence for such a claim continues to be deeply unclear to me" (2005, 231).

The idea of the specious present is further developed in the contemporary literature by Barry Dainton in what he calls the "Extensional Model". Like any model of temporal experience that relies on the idea of minimal extensive units, this model must involve two different levels of analysis: that of each experiential unit and that of the relation among successive units. Based on its two basic assumptions—that experience cannot be confined to 'an instant' and that "*every* temporally extended experience is an experience of succession" (Dainton 2000, 180)—the extensional model transforms the two aspects of the general conceptualist difficulty ("at once" and "in succession") into two requirements, each of which applies at a different level of analysis. While the "continuity requirement" asks how specious presents are combined "to form a phenomenally continuous stream of consciousness", the "dynamic requirement" asks how "change, succession, movement, persistence are all directly experienced over short intervals" (Dainton 2008, 370). The latter question concerns the dynamic nature of a single experiential unit. It is at this level of analysis that the implications of the intuitive description of continuity *as* transition come to the fore. In the final section, I address this issue in the context of James' and Whitehead's analyses of the intrinsic nature of a single experiential unit. The differences between them, as well as the similarities, will allow us to reexamine the gap between analysis and experience.

7.2 Russell on Continuity and Transition

7.2.1 *The Feeling of Transition and the Theory of Continuity*

Like James, Russell employs the idea of the specious present in analysing temporal experience in his *Theory of Knowledge*.[3] He also agrees with James that the immediate experience of the specious present is required for the experience of time as a whole. However, contrary to James, Russell does not believe that a clarification of the meaning and role of the intuitive concept of transition is required. The reason becomes clear in Russell's *Our Knowledge of the External World* (1914). Although he admits that "the events which we experience have not only a finite duration, but a duration which cannot sink below a certain minimum" (1914, 126), he still insists that our experience does not reveal some metaphysical truth about the nature of time and of transition or passage but rather its features. In other words, it is unnecessary to explain temporal continuity on the basis of *our experience of it as transition*. The concept of continuity is fully analysed in terms of its mathematical definition as "a property only possible to a *series* of terms [...] so that we can say of any two that one comes *before* the other" (1914, 137). This does not mean, however, that a thing suddenly jumps from one position to another since as shown by the infinite divisibility of time (and distance) there is never a next position or a next instant (1914, 140).

Of course, Russell is aware of the discomfort created by getting rid of the intuitive identification of continuity and felt transition. Referring to his critics' claim that there seems to be an unbridgeable gap between the mathematical definition of continuity and our experience of it *as* transition he writes:

> [...] there remains a feeling—of the kind that led Zeno to the contention that the arrow in its flight is at rest—which suggests that points and instants, even if they are infinitely numerous, can only give a jerky motion, a succession of different immobilities, not the *smooth transition* with which the senses have made us familiar (1914, 136; emphasis added).

Russell insists however that there must be a way to avoid this uneasiness. For example, our physiology can probably explain the gap, thus at least proving that the mathematical model can be applied to the physical world. Therefore, in spite of the fact that continuity is "easier to feel than to define", as he admits, our feeling of it need not be taken into account in the analysis (1914, 135). In fact, it is better for us to eliminate our discomfort by means of a conscious effort to *feel* the nature of the mathematical theory of continuity and the concept of a series upon which it is based. This is difficult simply because when a theory is comprehended logically, there is often "a long and serious labour still required in order to *feel* it: it is necessary to dwell upon it [...] to acquire the kind of intimacy which, in the case of a foreign language, would enable us to think and dream in it" (1914, 136).

---

[3] Russell defines the specious present of 'a momentary total experience' (i.e., of group of experiences whose objects are experienced together) as "the period of time within which an object must lie in order to be a sense-datum in that experience" (1913, 68).

The idea of dwelling on the mathematical analysis in order to feel it relates to what Russell refers to as "the psychological argument" which states that the mathematical description is consistent with our feeling of change and motion and should be able to explain it. To be sure, Russell is not implying that mathematical moments are part of sense data; in fact, he argues (as mentioned above), that even if the mathematical theory is adequate, there is obviously nothing momentary in the sensorial domain. Nonetheless, Russell insists that even though our temporal experience cannot be fully expressed by the intellect it should nevertheless be analysed using it. Thus, ideas such as points and instants are all logical constructions that are built upon sense data and can be legitimately ascribed to the world. In other words, the reference from the sensorial realm to the physical world is valid. Russell, who wishes to establish the correlation between the mathematical explanation and our senses, argues that in spite of the gap between these two realms (which is only due to our inability to imagine the idea of the series), what actually happens must have a mathematical structure.[4] Motion, for example, must exemplify mathematical continuity and the mathematical theory, for its part, must explain our feeling of motion as smooth transition:

> It will therefore be desirable to consider explicitly the mathematical account of motion, with a view to making its logical possibility *felt*. The mathematical account of motion is perhaps artificially simplified when regarded as describing what actually occurs in the physical world; but what actually occurs must be capable, by a certain amount of logical manipulation, of being brought within the scope of the mathematical account. (Russell 1914, 139; emphasis added)

The controversy between Russell and his opponents, which include James and Whitehead, ultimately centres on the role of our crude feeling of transition within the conceptualization of temporal processes such as motion.[5] While Russell denies the relevance of the intuitive definition of continuity, James and Whitehead deny the

[4] Russell's attempt to ground the theory in our experience of sense data motivated the criticism by Adolf Grünbaum. Grünbaum claims that "we can now offer a reason for Russell's inability in *Our Knowledge of the External World* to provide a justification of his assumption that the temporal order of physical events is that of the Cantorean continuum: The program of logical construction which he endorses in that work involves epistemological commitments which preclude such a justification. Whitehead sees this fact and accepts the consequences of his sensationist epistemological commitments by developing a *pulsational* theory of becoming. While following Whitehead by enunciating his Maxim of Logical Construction, Russell gratuitously and falsely assumes that a Cantorean theory of becoming can be made intelligible and convincing on the basis of the *sensationist* assumptions of his program" (1950, 186).

[5] The dispute usually takes place in the context of Russell's 'at-at theory' of motion which implies that "nothing happens when a body moves except that it is in different places at different times" (1914, 144). For example, Carol Cleland argues that Russell's conception "does not fit well with our intuitions about change" (1990, 257). John Carroll argues that "the Russellian view does not adequately characterize our common sense concept of motion at an instant" (2002, 50). And Graham Priest states that: "this conception of motion jars against our intuitive notion of motion as a genuine flux" (1985, 340). At the opening of his paper, "Some Cosmological Implications of Temporal Experience", which appears in this volume, Dainton also expresses this intuitive discomfort.

applicability of the mathematical analysis of continuity to our experience and thus to the facts of concrete reality. A temporal process must involve *more* than the mathematical conception of order. As James writes, "the mathematical definition of continuous quantity as 'that between any two elements or terms of which there is another term,' is directly opposed to the more empirical or perceptual notion that anything is continuous when its parts appear as immediate next neighbors, with absolutely nothing between" (1911, 95). In fact, James finds this line of thought so unreasonable that he notes: "It is probable that Russell's denial of change, etc., is meant to apply only to the mathematical world. It would be unfair to charge him with writing metaphysics in these passages" (1911, 94). James' rejection of the mathematical analysis can be further clarified on the basis of his distinction between two kinds of infinity[6]: the 'standing' infinity which is applied to space, past time and existing beings and the 'growing' infinity which is applied to processes in the making, that is, to motion, change and activity. According to James, Russell's solution bypasses the real difficulty, which relates to the growing type of infinity and may be called *physical* since it concerns the process of completing the route itself. Even if a process can be analysed in terms of the standing type of infinity after being completed, the analysis should not be understood as a transcription of what brought the process into existence. Whitehead's similar objection to the applicability of the mathematical analysis leads him to his most oft-quoted theory regarding the construction of continuity, i.e., his epochal theory of time, which states that "every act of becoming must have an immediate successor, if we admit that something becomes" (1929, 69) and implies that continuity is something which becomes while becoming is not itself continuous.[7] Thus, James and Whitehead both object not to the mathematical account of continuity itself but rather to what James dubbed the "arithmetization" of intuition (1911, 94).

7.2.2 The Two Senses in Which Russell Denies Transition: Change in Intrinsic Properties and Change as an Intrinsic Property

Against this background we can now address Russell's specific arguments concerning the two aspects that our intuitive characterization of the idea of transition seems to involve: first, there is the idea of transition as indicating a genuine connectivity, or internal relatedness, among the parts of a temporal process and second, there is

[6]The subject of infinity arises in the context of the debate between "the theory of discontinuity", which understands time as growing "by finite buds or drops", and the rival analysis of time as a continuous magnitude that can be divided to infinity (1911, 80).

[7]Whitehead writes: "There is becoming of continuity, but no continuity of becoming. The actual occasions are the creatures which become, and they constitute a continuously extensive world. In other words, extensiveness becomes, but 'becoming' is not itself extensive" (1929, 35).

the idea of transition itself as a dynamic feature, or in other words an intrinsic property, required for the reality of change in general.

Russell's denial of genuine connectivity among the phases of a temporal process is consistent with his overall rejection of internal relations, i.e. the idea that "every relation is grounded in the natures of the related terms" as he formulates the idea in his criticism of Bradley (1906–1907, 37).[8] The refutation of the idea of an all-embracing wholeness (and the idea of internal relatedness upon which it is based) reappears in Russell's examination of Bergson's philosophy.[9] In this case, it is the connectivity between the phases of a temporal process, i.e. the "interpenetration of past and present" underlying Bergson's conception of duration (Russell 1912, 338), that is being rejected. When referring to Bergson's characterization of temporal processes as undivided wholes, Russell writes that "this is part of a much more general doctrine, which holds that analysis always falsifies, because the parts of a complex whole are different, as combined in that whole, from what they would otherwise be" (1914, 157). Russell argues, in contrast, that temporal processes only imply relations; they are not made out of their parts or, as he puts it, motion is not made out of motions (1912, 341). Moreover, temporal relations themselves are not internal and therefore cannot somehow affect the nature of an object (1959, 59). In fact, temporal relations are not relations between events, "but only between moments" (1903, 482; also see 1901, 47) and, as Hager explains, since "points (or instants) lack complexity and do not differ in respect of intrinsic properties, the relation between any pair of points (or instants) is evidently an external relation *par excellence*" (1994, 138). Therefore, it is meaningless to claim that there is a real connectivity between the phases of the temporal process or that its parts are connected in such a way that they cannot be analysed separately. As Russell explains in *Principles of Mathematics*, this misconception is a result of confusing a thing with the change it undergoes, that is, with its external relations (1903, 478). In *Our Knowledge of the External World*, he further claims that if we take temporal relations to be a formative part of what a thing is, i.e. if we think that it is defined through its relations, then we cannot avoid the conclusion that there can be just one fact concerning that thing. In other words, if a thing is different in any one of its relations, then two facts indicate that there are two things (1914, 157). Russell writes: In fact, we may give a precise statement of the doctrine we are combating in the form: *There can never be two facts concerning the same thing*. A fact concerning a thing always is or involves a relation to one or more entities; thus two facts concerning the same thing would involve two relations of the same thing. But the doctrine in question holds that a thing is so modified by its relations that it cannot

[8] Russell reads Bradley as if he endorsed "the axiom of internal relations". However, the question of whether Bradley himself admitted internal relations or rejected relations on the whole is still controversial. For example, Peter Hylton, who mentions Bradley's tenet that all relations are unreal, further emphasizes the inability of internal relations to attain the non-relational unity (1990, 54–55).

[9] See Russell 1912.

be the same in one relation as in another. Hence, if this doctrine is true, there can never be more than one fact concerning any one thing (1914, 157).

Russell's doctrine of change as merely a relational difference is fully expressed when we consider its concomitant aspect. Thus, it is not simply that a thing cannot change in relation to its intrinsic properties, but rather that change itself is not an intrinsic property and the conception of transition as the mark of *action* at the fundamental level of reality is denied. "There is no such thing as a state of change" Russell writes, and criticizes Leibniz's assumption that "every body must contain in itself a principle of change, i.e. force or activity, by means of which a meaning is given to a *state* of change" (1903, xxi; 1900, 98). Russell's at-at theory of motion, which informs his analysis of Zeno's flying arrow paradox, clarifies this point.[10] Russell suggests that the power of this paradox resides in Zeno's implied assumption that when a thing is in a process of change, there must be some internal state of change in that thing (1912, 340). By showing that there is no such state, Zeno denies the reality of change and motion. But Russell claims that a state of motion is not required for motion as a whole and that, in more general terms, a state of change is unnecessary for the reality of change. Motion, like any other kind of change, merely implies a relational difference and does not involve a "transition from place to place" (1903, 480). Of course, a moving body "always passes by a gradual transition" if by this we are using the term 'transition' in the mathematical sense, such that "no two terms are consecutive, but between any two there are others" (1914, 142; 138). There is, however, no transition in the sense of some aspect that is required for the analysis of temporal continuity and which indicates that the parts of the temporal process are bound together.

7.3 Transition and Connectivity

With the implications of Russell's denial of change in mind, we can revisit James' and Whitehead's descriptions of the continuous transition between successive elements and within each element. As we have seen, the relations among successive units in James' description are based on the idea of "immediate next neighbors, with absolutely nothing between." After all, according to James the *absence* of gaps is precisely what we mean by 'continuity', so that "continuity seems thus to have a primarily negative meaning", and objects are continuous with each other "when they are represented as mutually next, adjacent, or contiguous" (1988a, 32). Also, for Whitehead, "A purely temporal nexus of occasions is continuous when, with the exception of the earliest and the latest occasions, each occasion is contiguous with an earlier occasion and a later occasion. The nexus will then form an unbroken thread in temporal or serial order" (1933, 202). These statements suggest a similarity to Aristotle's account of continuity, which is also framed in terms of contiguity.[11]

[10] See footnote 5.

[11] *Physics* V. 3 and VI. 1.

Indeed, the similarity goes even further since for Aristotle continuity implies more than contiguity. Thus, while contiguity involves succession and contact, two successive things are *continuous* when their limits become identical, i.e., when they are fused into one. Thus, continuity forms unity. James and Whitehead, like Aristotle, suggest a physical, rather than mathematical, account of continuity. Unfortunately, James misses the similarity when he identifies adjacency, i.e. "the lowest grade of union", with Aristotle's continuity in his *Manuscripts* (1988b, 324).

The next question to be tackled is *how* these successive units are combined to create a unified stream. One way of looking at this process is in terms of internal relations, which are—if we follow Moore's characterization—relations essential to their relata. In this case, the essential relations between temporal pulses connect them in such a way that allows for continuity. Whitehead's description of the connectivity among actual entities is based, in part, on this line of thought. Thus, the relations among these units of composition are internal, that is, "they are constitutive of what the event is in itself" (1926, 106). But what about Russell's objection that a thing cannot be characterized by its relations since its identity is not preserved?

In this case, James' and Whitehead's responses diverge. Whitehead's answer is that every drop of temporal extension becomes like a complete whole, with an internal set of relations to other drops. It is a discrete pulse of experience which (in contrast to the traditional characterization of the eternal substance) becomes and perishes and contains only *one* set of relations. Temporal extension is then understood as a structure (or a 'nexus' as Whitehead calls it in, for example, the above quotation from *Adventures of Ideas*), which is a result of the binding of such entities through internal relations (1929, 288). In fact, Whitehead also admits external relations among entities and presents a systematic method of analysing them.[12] However, taken on its own, the analysis of connectivity in terms of external relations is not inclusive since it does not clarify the characterization of temporal continuity *as* transition, i.e., it does not explain how atomic individual bricks create the experience of a continuous flow. The importance of external connections is again *physical*, since they serve as the foundation that allows transition among entities: "Thus the notion of continuous transmission in science must be replaced by the notion of immediate transmission through a route of successive quanta of extensiveness" (Whitehead 1929, 307). Transition thus becomes the mark of connectivity between successive entities. One consequence of Whitehead's idea of an entity as a complete set of relations is that it cannot change internally. Whitehead argues: "Locke misses one essential doctrine, namely, that the doctrine of internal relations makes it impossible to attribute 'change' to any actual entity. Every actual entity is what it is, and is with its definite status in the universe, determined by its internal relations to other actual entities" (1929, 58–59). It is precisely because of the idea of internal relations, and the elimination of the possibility of entities changing, that 'transition' is understood first and foremost as the mark of connectivity between successive units.

[12] These are the connections between what Whitehead refers to as the extensive aspect of the entities, i.e., their region.

Continuity thus implies the process of transition *from* one drop to its successor and is the reason why Whitehead eventually suggests that we understand succession through causation, which is its *reason*. Our perception of extensive relations is always accompanied by a direct perception of causation, which represents the asymmetrical temporal order of succession and ensures our experience of passage and flow. In fact, the extensive continuum is not a mathematical continuum explainable in terms of an infinite series of points and instants, but rather points and instants represent possibilities for division that are abstracted from the real connectivity between extensive drops. In other words, 'moments' are abstracted from concrete relatedness.

James of course shares this view of the priority of concrete experience: "The pragmatist himself has no objection to abstractions" he writes, "but he never ascribes to them a higher grade of reality" (1908, 11). However, James also shares Russell's opposition to the idea of internal relations, which entails "the individual absolute, with its parts co-implicated through and through, so that there is nothing in any part by which any other part can remain inwardly unaffected" (James 1909, 68). However, James' rejection of internal relations and his criticism of Bradley's monism do not imply the externality of relations. The key to understanding James' position is the word 'some': "Radical empiricism and pluralism stand out for the legitimacy of the notion of *some*: each part of the world is in some ways connected, in some other ways not connected with its other parts" (1909, 79). It is on this basis that James argues in "The Thing and its Relations" (1905) that we *can* actually attribute two or more relations to the same thing, as our experience shows, since what actually happens appears to stand *in between* those things that come before and after it. This does not imply that an object appears twice, for even "if I 'took' it a thousand times I should still *see* it as a unit. Its unity is aboriginal, just as, in my successive takings of it, the multiplicity is aboriginal" (1905, 34). James is claiming that our inability to capture, in one phrase, the variety of relations that characterizes a thing is due only to a *verbal* difficulty. In contrast to Russell, James' criticism of the monistic belief (which "allows of no taking up and dropping of connexions, for in the all the parts are essentially and eternally co-implicated" (1909, 324)) is designed to establish a "strung along type" of unity. Thus, James is implying that the experienced continuity, or 'concatenation', is a continuous flowing type of unity in which influence between components does not eliminate the identity of each such component (1909, 356–357).[13] This idea had already been elegantly expressed in the *Principles*, in which James refers to the experience of hearing thunder as follows:

> Into the awareness of the thunder itself the awareness of the previous silence creeps and continues; for what we hear when the thunder crashes is not thunder *pure*, but thunder-breaking-upon-silence-and-contrasting-with-it. Our feeling of the same objective thunder, coming in this way, is quite different from what it would be were the thunder a continuation of previous thunder. The thunder itself we believe to abolish and exclude the silence; but the

[13] Gale writes: "these immediate neighbors are identical but not THAT identical" (1997, 174). See also Skrupskelis (1988, xxxviii).

> *feeling* of the thunder is also a feeling of the silence as just gone; and it would be difficult to find in the actual concrete consciousness of man a feeling so limited to the present as not to have an inkling of anything that went before. (1890, 240–241)

Real connectivity thus implies the *intermingling* of successive phases and this is the basis of our feeling of "*co-conscious transition*" as James calls it (1904, 536). Our experience is "without a breach, crack, or division" since the units of composition *are not distinct* and "all real units of experience *overlap*" (1909, 287).

James and Whitehead not only reject the applicability of the mathematical theory of continuity to experience and instead adopt the necessary absence of a gap (by contiguity or overlapping) but also consider this structural aspect to be the platform for *real* connectivity, which involves substantial influence and relates to the *physical* aspect of the construction of continuity. On their own, succession, contiguity and overlapping are each insufficient for this purpose. Thus, in order to provide an explanation of our experience of succession *as* a flow, Whitehead reconceptualises succession in terms of causation and James (following Bergson) rejects the idea of "abstract succession" (1909, 261) and turns directly to the immediacy of experienced concatenation. "The transition *is* the bridge, is the fact that the next thing has come" according to James (1988a, 34).

7.4 Transition and Action

The dynamic implication of the idea of transition becomes even more pronounced when we consider the internal structure of each experiential unit. We therefore return to the question of how the successive phases within each unit all belong to the scope of the present. We are able to resolve this difficulty, as James does, by reconceptualizing the 'present' itself. James writes: "The literally present moment is a purely verbal supposition, not a position" (1909, 254). In other words, the conceptual difficulty is, as its name implies, just conceptual and disappears when we understand 'the present' to be an experiential feature. James continues: "The only present ever realized concretely being the 'passing moment' in which the dying rearward of time and its dawning future forever mix their lights" (1909, 254).

Furthermore, because the *experienced* present, in contrast to the strict present, is extensive, its successive phases are not simultaneous: "[N]o two terms can possibly be *simultaneously* perceived to differ, unless, in a preliminary operation, we have successively attended to each, and, in so doing, had *the transitional sensation of difference* between them aroused" (1890, 495; emphasis added). Therefore, the only way to make sense of our immediate experience of continuous transition (which is the basis for the experience of temporality in general) is on the basis of a direct awareness of successive contents, i.e. the immediate perception of before and after. Notice that this sensation of difference (or 'feeling of relation' as James sometimes calls it) is experienced dynamically, due simply to the dynamic ingredient that is primarily immersed within it. In other words, the sensation of difference is *already dynamic*, i.e. it is the feeling of transition or flow in a certain direction. Russell explains temporal experience along the same lines. In *Theory of Knowledge*, he writes: "[S]uccession may be immediately experienced between parts of one sense datum [...] in this case, the two objects of which one is succeeded by the other are both parts of the present" (1913, 73). Nevertheless, a complete description of our immediate experience of relational difference must also involve a reference to its dynamic nature, which however cannot be further analysed and remains enigmatic even after a meticulous reading of Russell's suggestion.[14] Russell writes:

> In the understanding of the abstract "before", which is what we are trying to isolate, there must be some kind of reference to terms, something, in fact, which we call "sense" or "direction". The two propositions "A is before B" and "B is before A" contain the same constituents, and they are put together according to the same form; thus the difference is neither in the form nor in the constituents. It would thus seem that a relation must have essentially some "from-and-to" character, even in its most abstract form [...] From what has been said, it follows that such words as *before* and *after*, *greater* and *less*, and so on, are not the names of relations: they always involve, in addition to the relation, an indication as to "sense". (Russell 1913, 86, 88)

Nonetheless, shifting the focus to the phenomenological meaning of the 'present' does not confront the question of how to conceptualize the experiential togetherness of earlier and later phases. Of course, this is how experience is *felt*, but how, if at all, can this feeling be accounted for?

Whitehead, who attempts to clarify every element of our experience within his general metaphysical framework, must try to answer this question. Thus, Whitehead handles the conceptual difficulty at this ultimate level by treating the becoming of an entity in an atemporal manner, namely as something which "is not in physical time" and does not involve a temporal succession (1929, 283). In fact, the relation between physical time and the becoming of an entity, i.e. concrescence, works in the opposite direction, such that it is the becoming of entities that *underlies* physical time.[15] Thus, Whitehead distinguishes between two different modes of analysis. While the first focuses on the nature of each and every drop as an individual quantum

[14]This strategy is controversial because it simply transforms the problem of how to explain the dynamic nature of temporal experience in general to the level of its ultimate constituents. Ian Phillips raises this problem in relation to Dainton's similar scheme. According to Dainton, the dynamic nature of the specious present stems from the fact that its contents "possess an internal temporal organization, an intrinsic and directed animation" (Dainton 2000, 176). In other words, it is not enough that the successive phases are parts of "a single (extended) experience that is *sensed as a whole*"; rather, the contents of one specious present are "truly dynamic", i.e., they possess the character "of a *ball moving and falling*" (Dainton 2008, 370). He also claims that "these contents have the form of temporal fields or spreads of content possessing inherent directedness or 'flow' (e.g. a ball moving to the right)" (Dainton 2003). In response, Phillips makes the following complaint: "[...] at this point, Dainton directly appeals to irreducibly temporal properties—the flow of experience itself. But, if it is legitimate to appeal to irreducibly temporal properties in the context of explaining our perception of order and flow, why not simply do the same with respect to our perception of temporal properties more generally?" (2010, 189)

[15]The same structure is implied by the Leibnizian system in which the internal action of the monads is prior to temporal extension and in fact constitutes it.

(rather than on its relations), the second focuses on the extensive character of the drops (rather than on their individual character). It is the second mode of analysis that explains physical time and succession, both of which are constituted by the relations among drops, as was explained above. There is indeed a sense in which the becoming of an entity, i.e. concrescence, can be viewed as a succession of subsequent phases (1929, 220). But even though "there is a growth from phase to phase", each phase "presupposes the entire quantum" (1929, 283). Thus, "the process of self-creation of an actual entity is not a process in time; it is, rather, an atemporal process leading to the momentary appearance of the completed actual entity in space time" (Malin 2009, 65). Thus, Sherburn argues that: "concrescence is not in time; rather, time is in concrescence in the sense of being an abstraction from actual entities [...] then it seems strange to talk about one phase of concrescence as prior to another when the passage from phase to phase is not in physical time" (1981, 38). Therefore, and this is the crucial point, although we must *describe* each pulse of change in terms of its parts and the relations among them, each pulse is, in and of itself, prior to its parts. Whitehead thus abandons the characterization of each unit in terms of multiplicity. *Each unit is an undivided pulse of a primordial type of transition, which is the reason for the flow of time and for relational changes.*

With Whitehead's analysis in mind, we return to James who, as mentioned, responds to the problem of the common membership (or "experiential togetherness" in Whitehead's words) of the earlier and later phases within the units primarily by highlighting the limits of conceptual reasoning. Language and logic are incapable of expressing what actually happens, such that "The concept 'many' is not the concept 'one'; therefore the manyness-in-oneness which perception offers is impossible to construe intellectually" (1911, 51). Nevertheless, alongside his appeal to "turn our backs upon our winged concepts altogether, and bury ourselves in the thickness of those passing moments over the surface of which they fly" (1909, 251–252), James repeatedly attempts to achieve a clearer grasp of the idea of an ultimate unit of experience. In fact, the ways in which he sometimes portrays this idea have led commentators (e.g., Sprigge 1993, 221) to associate his analysis of time with that of Whitehead. James' most cited statement in this context appears in *Some Problems of Philosophy*: "Either your experience is of no content, of no change, or it is of a perceptible amount of content or change. Your acquaintance with reality grows literally by buds or drops of perception. Intellectually and on reflection you can divide these into components, but as immediately given, they come totally or not at all" (1911, 80). Furthermore, while discussing Zeno's paradoxes in *A Pluralistic Universe*, he claims that "time itself comes in drops" (1909, 232) and explains:

> If in the natural world there were no other way of getting things save by such successive addition of their logically involved fractions, no complete units or whole things would ever come into being [...] But in point of fact, nature does n't make eggs by making first half an egg, then a quarter, then an eighth [...] She either makes a whole egg at once or none at all, and so of all her other units [...] If all change went thus drop-wise, so to speak, if real time sprouted or grew by units of duration of determinate amount, just as our perception of it grow by pulses, there would be no zenonian paradoxes. (1909, 230–231)

A closer look at James' statements regarding the inner nature of the pulses reveals the distance between him and Whitehead. In particular, it seems that by referring to the idea of "units of amount bursting into being 'at a stroke'" (James 1911, 80), which Whitehead describes as "not in physical time", James simply highlights the *unity* of the pulses, i.e., their indivisibility. His usage of "pulses", since it remains intimately connected to experience, does not have the flavour of atemporal becoming (nor does it indicate the separation between the pulses).[16] In fact, James sometimes treats this wholeness in temporal terms. He quotes Hodgson who claims that "a former and a latter are included in the minimum of consciousness" and justifies this by claiming that "if we do not feel both past and present in one field of feeling, we feel them not at all" (1890, 607; 1909, 283). Furthermore, in his appendix to *The Many and the One* (1903–1909) he contemplates the essential extensive character of the pulses (without which reality wouldn't grow) and considers the Bergsonian position that "the units that actually come have the many-in-one form and are undivided in act, and divisible only *after the fact*" (1988a, 45). The idea of a pulse simply expresses the aboriginal and unanalysable unity of each unit of experience. James writes: "Here, then, inside of the minimal pulses of experience, is realized that very inner complexity which the transcendentalists say only the absolute can genuinely possess. The gist of the matter is always the same—something ever goes indissolubly with something else. You cannot separate the same from its other, except by abandoning the real altogether and taking to the conceptual system" (1909, 284).

This leads us to the common element in James' and Whitehead's suggestions. For both of them, succession is ultimately only a second-order conception, an abstraction from the concrete facts of experience. At this level, the concept of transition is approached via unity, like some primitive type of 'action' or 'becoming' that cannot be further analysed. This is the same idea expressed by James in the *Principles* when he argues that a transitive thought cannot be divided:

> Let anyone try to cut a thought across in the middle and get a look at its section, and he will see how difficult the introspective observation of the transitive tracts is. The rush of the thought is so headlong that it almost always brings us up at the conclusion before we can arrest it. Or if our purpose is nimble enough and we do arrest it, it ceases forthwith to be itself. (1890, 244)

In spite of the fact that what has become is subject to *analysis* in terms of earlier and later phases when each experiential drop is considered in and of itself, according to both James and Whitehead, it appears all at once as an amount of a felt

[16] A related discussion concerns the tension between James' characterization of the pulses as 'discrete' on the one hand (a characterization that indicates the separation between them) and his Bergsonian description of intermingling elements on the other. It has been recognized that James pulls in these two different directions. For example, Hare notes in his introduction to *Some Problems of Philosophy*, that in spite of the fact that James abandons the idea of discrete change as "the whole story of novelty", he does not give it up completely; thus his task is to "work out a theory that recognizes both discreteness and continuity in growth" (1979, xxx). For a detailed discussion of this tension, see also Gale (1997).

transition. In principle, this felt transition does not result from the relations among the successive phases that constitute each drop since the phases are not actual parts but only the outcome of a *methodological distinction*. It is only through discriminative attention that we can recognize them. As Lowe explains, not only is the earlier-later relation directly provided to us in our experience, but "we [also] know the meaning of the earlier-later relation in virtue of an intuition of the passage of time in itself" (1941, 117). The claim is that temporality is already immersed within our experience of a relational difference and that the immediate givenness of successive phases inevitably brings us back to the passage of time itself. Transition is being understood here *as* an intrinsic property—as a dynamic feature—which is the reason for, rather than the result of, our experience of temporal processes as a whole. Ultimately, continuity cannot be explained in terms of succession or of co-existence. Both are abstractions that cannot capture the essence of time as being "what hinders everything from being given at once" (Bergson 1946, 93).

Bibliography

Aristotle. 1999. *Physics*. Trans. R. Waterfield and D. Bostock (ed.). Oxford: Oxford University Press.

Bergson, H. 1946. *The creative mind*. New York: The Philosophical Library.

Blake, R.M. 1926. The paradox of temporal process. *The Journal of Philosophy* 23(24): 645–654.

Carroll, J.W. 2002. Instantaneous motion. *Philosophical Studies* 110: 49–67.

Cleland, C. 1990. The difference between real change and mere Cambridge change. *Philosophical Studies* 60: 257–280.

Dainton, B. 2000. *Stream of consciousness: Unity and continuity in conscious experience*. London: Routledge.

Dainton, B. 2003. Time in experience: Reply to Gallagher. *Psyche* 9(12). http://barrydainton.com/wp-content/uploads/2012/09/Reply-to-Gallagher-v2.pdf

Dainton, B. 2008. Sensing change. *Philosophical Issues* 18(1): 364–384.

Gale, R. 1997. From the specious to the suspicious present: The Jack Horner phenomenology of William James. *The Journal of Speculative Philosophy* 11(3): 163–189.

Grünbaum, A. 1950. Relativity and the atomicity of becoming. *The Review of Metaphysics* 4(2): 143–186.

Hager, J.P. 1994. *Continuity and change in the development of Russell's philosophy*. Dordrecht: Kluwer Academic Publishers.

Hare, H.P. 1979. Introduction. In *Some problems of philosophy*, ed. W. James, xiii–xli. Cambridge/London: Harvard University Press.

Hylton, P. 1990. *Russell, idealism, and the emergence of analytic philosophy*. Oxford/New York: Clarendon.

James, W. 1890. *The principles of psychology*. New York: Dover Publications.

James, W. 1904. A world of pure experience. *The Journal of Philosophy, Psychology and Scientific Methods* 1(20): 533–543.

James, W. 1905. The thing and its relations. *The Journal of Philosophy, Psychology and Scientific Methods* 2(2): 29–41.

James, W. 1908. The pragmatist account of truth and its misunderstanders. *The Philosophical Review* 17(1): 1–17.

James, W. 1909/1996. *A pluralistic universe*. Lincoln/London: University of Nebraska Press.

James, W. 1911/1979. *Some problems of philosophy*. Cambridge/London: Harvard University Press.
James, W. 1988a. *Manuscript essays and notes*. Cambridge/London: Harvard University Press.
James, W. 1988b. *Manuscript lectures*. Cambridge/London: Harvard University Press.
Kelly, S. 2005. Temporal awareness. In *Phenomenology and philosophy of mind*, ed. D. Woodruff Smith and A.L. Thomasson, 222–234. Oxford: Oxford University Press.
Lowe, V. 1941. William James and Whitehead's doctrine of prehensions. *The Journal of Philosophy* 38(5): 113–126.
Malin, S. 2009. Whitehead's philosophy and Quantum Mechanics (QM): A tribute to Abner Shimony. In *Quantum reality, relativistic causality, and closing the epistemic circle: Essays in honour of Abner Shimony*, ed. W.C. Myrvold and J. Christian, 63–68. Berlin and Heidelberg: Springer-Verlag.
Phillips, I. 2010. Perceiving temporal properties. *European Journal of Philosophy* 18(2): 176–202.
Priest, G. 1985. Inconsistencies in motion. *American Philosophical Quarterly* 22: 339–346.
Russell, B. 1900/2005. *The philosophy of Leibniz*. London/New York: Routledge.
Russell, B. 1901. On the notion of order. *Mind* 10(37): 30–51.
Russell, B. 1903/2010. *Principles of mathematics*. London/New York: Routledge.
Russell, B. 1906–1907. On the nature of truth. *Proceedings of the Aristotelian Society* 7: 28–49.
Russell, B. 1912. The philosophy of Bergson. *The Monist* 22(3): 321–347.
Russell, B. 1913/1992. *Theory of knowledge*. London/New York: Routledge.
Russell, B. 1914/2000. *Our knowledge of the external world*. London/New York: Routledge.
Russell, B. 1959. *My philosophical development*. New York: Simon & Schuster.
Sherburn, D. W. 1981. *A key to Whitehead's Process and reality*. Chicago: University of Chicago Press.
Skrupskelis, I.K. 1988. Introduction. In *Manuscript essays and notes*, ed. W. James, xiii–xlviii. Cambridge/London: Harvard University Press.
Sprigge, T. 1993. *James and Bradley: American truth and British reality*. Chicago/Illinois: Open Court.
Whitehead, A.N. 1920/1978. *Concept of nature*. Cambridge: Cambridge University Press.
Whitehead, A.N. 1922. *The principle of relativity*. London: Cambridge University Press.
Whitehead, A.N. 1926/1953. *Science and the modern world*. London: Cambridge University Press.
Whitehead, A.N. 1929/1978. *Process and reality*. New York: The Free Press.
Whitehead, A.N. 1933/1967. *Adventures of ideas*. New York: The Free Press.

Chapter 8
Consciousness and the Present

Ulrich Meyer

Abstract A perennial question in the philosophy of time concerns the relation between the objective "physical time" that features in empirical theories of motion and the subjective "human time" in which our own experiences unfold. This article is about one facet of this broader question: whether the phenomenon of consciousness allows us to make a principled distinction between the present and other times. A number of authors have argued that, without conscious observers, there would be no distinctions of past, present, and future. This paper defends the opposing thesis that there is no interesting connection between consciousness and presentness.

Keywords Time and consciousness • Open future • Physical time • Human time • Presentism • Eternalism • Growing block

8.1

Amongst the many questions that philosophers of time worry about, two issues have turned out to be particularly controversial. The first issue concerns the relationship between the *physical time* that features in scientific theories and the *human time* in which our own conscious experiences unfold. Philosophers in the continental tradition often take the latter to be the more fundamental notion, and argue that we need to explain physical time in terms of human time. Analytic philosophers tend to adopt the opposite approach and try to account for the temporal dimension of our subjective experiences in terms of objective physical time. But these divisions are far from universal. For example, Bertrand Russell (1915a: 116) argues that we need to construct instants of time out of immediate sense awareness, and both Alfred North Whitehead (1919) and Michael Dummett (2000, 2005) express

U. Meyer (✉)
Department of Philosophy, Colgate University, Hamilton, NY 13346, USA
e-mail: umeyer@colgate.edu

Y. Dolev, M. Roubach (eds.), *Cosmological and Psychological Time*, Boston Studies in the Philosophy and History of Science 285,
DOI 10.1007/978-3-319-22590-6_8

dissatisfaction with the physicist's view that the time-series forms a continuum of extensionless instants.[1]

The second issue concerns the status of the present moment. *Eternalists* think that all times are alike, but others argue that there is a principled metaphysical difference between the present and past and future times: *presentists* hold that only present objects exist, but no past or future ones; and the *growing block view* contends that past and present are real, but not the future. There are also philosophers who think that there is a non-trivial sense in which time flows. In this case, the present would not be ontologically privileged, but rather mark the position in the time series to which the flow of time has advanced.

The two issues are not independent. If we privilege physical time over human time then we are likely to end up with a view on which there is nothing special about the present moment. Indeed, many authors think that modern physics flatly contradicts both presentism and the growing block view. According to the theory of relativity, there are no absolute facts about whether two spatially separated events e_1 and e_2 occur simultaneously. In some reference frames, e_1 and e_2 occur simultaneously, but there are also reference frames in which e_1 occurs before e_2, and reference frames in which e_2 occurs before e_1. But if there are no frame-independent facts about which events are present at the same time, we are told, then there cannot be any principled metaphysical difference between the present and other times. There is a lively debate about whether this is correct, and whether the theory of relativity does indeed rule out a principled distinction between the present and other times. But even if our best physics does not prevent us from assigning a privileged status to the present, it surely does not encourage such views, either. Physical theories treat all times the same and do not even contain terms like 'past', 'present', or 'future'.

On the other hand, it is not clear that we ought to privilege physical time in this way. Physics assumes that the time series is isomorphic to the real numbers, and this is clearly not how time is given to us in conscious experience. A prominent example is what William James (1890: 628) calls the *specious present*. In observing a hand move around the face of a clock, there is a phenomenal difference between seeing the hand first at one position and then at another, by looking twice, and *seeing it move*, by looking only once. Instead of a succession of different locations, we often perceive motion itself. The question is whether we should take this seriously, and conclude that instants of time are not pointlike. If we do, then physics would be mistaken about the basic structure of time.

Those who privilege physical time over human time usually try to account for the specious present in terms of a retention mechanism by means of which our brain synthesizes sensory information received over a short period of time into a momentary impression of change.[2] This view treats the time series as a continuum of

---

[1] See also Meyer (2005). Dolev (2007) attempts to bridge the gap between continental and analytic approaches to the philosophy of time.

[2] See Dainton (2008, 2011), Hoerl (2009), Paul (2010), and Prosser (2012). A survey of the neurological basis of temporal processing can be found in Mauk and Buonomano (2004) and Buhusi and Meck (2005); see also Canales (2009).

extensionless points, but Yuval Dolev (2012) objects that it is too much at variance with the phenomenology of temporal perception to be plausible.

This is not the place to settle the debate about whether human or physical time is more fundamental. In this paper, I want to focus on a more limited question instead. If we take our cue from physics then we are bound to reject both the importance of human time and the specialness of the present. My question is whether there is room for an opposing view that takes human time seriously and uses it to mark the present as metaphysically special. Does the phenomenon of consciousness allow us to make a principled distinction between the present and other times?

8.2

Consciousness could mark the present in either of two ways. One possibility is that consciousness generates presentness and that there would be no distinctions of past, present, and future without the existence of conscious beings who engage in the right sort of conscious activity. The second possibility is that presentness brings about consciousness. In that case, presentness would be a necessary condition for the occurrence of conscious activity, rather than the other way around.

Lynne Rudder Baker recently defended a version of the first thesis. On her view, "an event's occurring now depends on someone's being judgmentally aware of it now" (2010: 32). If there were no judgmental awareness then there would be no presentness, Baker contends, but she does not endorse the stronger thesis that the existence of time itself requires consciousness. Even without conscious observers, she thinks that there would still be instants of time that are ordered by the earlier-than relation. There would be time, but no distinctions between past, present, and future. Rather than take human time as fundamental, Baker believes it to be merely complementary.

One advantage of this proposal is that it does not commit us to the thesis that physical time can be reduced to human time. In his discussion of Russell's proposal that we construct times out of immediate sense awareness, Adolf Grünbaum (1953) notes that sense perception does not allow us to discern times that are, say, 10^{-1000} s apart, even though in a continuum there are infinitely many instants between them. If we try to account for physical time in terms of human time we are bound to end up with too few instants to form a continuum, which would leave us in the uncomfortable position of having to reject well-established physical theories for purely philosophical reasons. Baker's proposal does not have this drawback.

On the other hand, Baker's view does have the bizarre consequence that there are times that were never present because no conscious observers existed then. Nobody was judgmentally aware of anything one billion years ago, even though there were plenty of events occurring at that time. Baker's view thus conflicts with the truism that what makes something a *past time*—rather than a future time, or something altogether different—is that it was once present. I would think that this is already sufficient reason for rejecting the view, but that is not a point that I to press here.

Instead, I want to highlight a problem that arises for any view that claims that consciousness generates presentness.

Suppose that π is a sentence that reports that somebody is judgmentally aware of some event. To describe what is true at various instants that make up the time series, let us use George Myro's (1986a, b) "true at" operator '|'. When saturated with a term t for a time on the left and a sentence φ on the right, this operator allows us to form an expression '$t \mid \varphi$' that claims that φ is true at time t. Baker's theory of presentness is then committed to the truth of the following biconditional:

$$\text{Time } t \text{ is } \textit{the present} \text{ if and only if } t \mid \neq \tag{8.1}$$

If π is true at t then somebody is judgmentally aware of some event at that time, making it present. Conversely, a time would not be present, on her view, unless judgmental awareness marks it as such, requiring π to be true at that time.

With this in place, let us suppose the time is t_0 and that I am judgmentally aware of my writing this sentence. Then somebody (namely me) is now judgmentally aware of some event (my writing) and π is true at time t_0:

$$t_0 \mid \neq \tag{8.2}$$

This is the expected result, for it merely confirms that we are now at the present time. However, by remembering my own experiences, I also know that I *was* judgmentally aware of some event in the past, such as my writing the first sentence of this paper. It is therefore also true at t_0 that somebody (namely me) *was* judgmentally aware of some event (writing the first sentence). In terms of the standard past tense operator 'P' ("it was the case that"), we can write this as:

$$t_0 \mid P\neq \tag{8.3}$$

By prefixing the operator 'P' to a sentence φ, we shift the time at which φ gets evaluated to an earlier time. A sentence '$P\varphi$' is true at time t just in case there is an earlier time t' at which the embedded sentence φ is true:

$$t \mid P\varphi \text{ if and only if there is a time } t' < t \text{ such that } t' \mid \varphi \tag{8.4}$$

For example, it is true now that it was raining just in case there is a past time at which it is raining. Applied to our case, this means that (8.3) entails the existence of a time t' before the present time t_0 such that π is true then. So there is a time t' earlier than the present such that somebody is judgmentally aware of some event at that time, $t' \mid \pi$. This result might have been expected as well, but if π is true at t' then Baker's thesis (8.1) about the connection between consciousness and presentness entails that the time t' is *also* present, and there cannot be two different times t_0 and t' that are both *the* present.

This *reductio* of Baker's proposal uses (8.4) to spell out tensed claims containing the operator 'P' in terms of untensed claims involving the earlier-than relation '<'

and the true-at operator '|'. One might object that this is bound to miss the point of the proposal, which can only be rendered in tensed terms. Only at one time is π *presently* true, one might point out; at all other times it merely *was* or *will be* true. However, the thesis that we are considering is that facts about consciousness—and not facts about the *presentness* of consciousness—determine which time is present. If we needed an independent way of making sense of the "presentness" of what is going on then it would be this presentness, and not consciousness, that makes a time present. The present time would be as much marked by the presentness of a thunderstorm as by the presentness of judgmental awareness. There would be nothing special about consciousness in singling out one time as the present one. So the fault does not lie with the truism (8.4), but with the thesis that judgmental awareness singles out one time as the present.

The same issue arises for any other way of spelling out the view that presentness generates consciousness. Baker's proposal is that someone's being judgmentally aware of an event makes a time present, but one could easily envisage theories that privilege a different feature of our conscious experiences. If we let π be a sentence that reports that the proposed present-making feature of consciousness obtains then we run into the very same difficulty as Baker's proposal. The occurrence of features of consciousness fails to distinguish the present from past and future times. At best, we get a contrast between times at which some people are conscious and those at which nobody is paying attention.

This is not to deny the obvious fact that there are salient phenomenal differences between past, present, and future. We are only directly aware of present events while our access to past events relies on memory and that to future ones on anticipation. Present events are vivid in a way in which past and future ones are clearly not. In his 1915 article, "On the Experience of Time," Bertrand Russell (1915: 212) defends a view that initially looks quite similar to the thesis that consciousness generates presentness:

> [P]ast, present, and future arise from time-relations of subject and object, while earlier and later arise from time-relations of object and object. In a world in which there was no experience there would be no past, present, or future, but there might well be earlier and later.

As emerges in the course of the paper, though, Russell is really interested in a different issue. By "present," he really means something like "present to the mind." His project is to figure out how these distinctions matter to our epistemology. If we consider our epistemic situation at a given time then the distinctions between past, present, and future do indeed matter, but this only shows that present events are privileged *at the present time*, and all events are privileged in exactly the same way when they are present. And if all times are "special" then none of them is.

8.3

Having considered the thesis that consciousness generates presentness, let me now turn to the converse claim that presentness brings about consciousness. This would still be a view on which consciousness "marks" the present, but it would require that we have already identified some principled difference between the present and other times that is not spelled out in terms of features of consciousness.

The main alternative to "consciousness theories" of presentness are accounts that try to draw the distinction between the present and other times in ontological terms. Since eternalism treats all times the same, it clearly does not support the view that presentness generates consciousness. The same is true for the presentist thesis that nothing exists that is not present. Perhaps there is some sense in which presentists could say that presentness generates consciousness, but that would be a trivial consequence of their view. If nothing exists unless it presently exists then there is nothing remarkable about the connection between presentness and *consciousness*. Presentness would be a necessary condition for consciousness, but it would also be a necessary condition for orange blossoms and tea cups and everything else that exists.

This leaves us with the growing block view, which contends that only past and present are real, but not the future. One proposal that explores this option for linking presentness and consciousness is Peter Forrest's *glowing* block view, which supplements the traditional growing block view with the claim that only beings at the edge of the growing block are conscious. His thesis is that consciousness is a product of the growing of the block, and that the past is a dead wasteland inhabited by unconscious zombies: "Life and sentience are, I submit, activities not states. Activities only occur on the boundary of reality, while states can be in the past. […] The past is […] dead" (2004: 359).

Forrest's view entails that presentness generates consciousness in a non-trivial sense. It also allows us to take seriously the intuition that there is something metaphysically remarkable about the vividness of our present experiences. Since the passage of time turns us into zombies, this vividness would *fade* once the experience is no longer present. Forrest could thus reject (8.3) and avoid the difficulties that beset Baker's theory.

Chris Heathwood (2005: 250) objects that the glowing block view would commit us to the implausible thesis that the sentence "Caesar was *conscious* when he crossed the Rubicon" has very different truth conditions than "Caesar was *wet* when he crossed the Rubicon." Forrest (2006) replies that this is just part of what makes consciousness special, which strikes me as a satisfactory response. Since it is Forrest's view that there is something very peculiar about the temporal aspect of consciousness, he should be allowed to spell out what this peculiarity consists in. This does not mean, however, that the proposal itself is plausible. In my view, the main problem with the glowing block view is what it says about time, not what it says about consciousness.

Growing blockers might deny the reality of the future, but they cannot reject *all* future tense claims as either false or meaningless. The growing block view itself is a thesis about the future, namely that the block will continue to grow and that there *will* be more to reality than there is right now. It is only this feature that distinguishes a *growing* block from an eternalist view that postulates the existence of a last time. Moreover, once reality has acquired the additional strata that the growing block view claims it will acquire, all the truths of logic *will* hold of the larger block in the same way in which they hold of the block in its current state. So there are truths about the future that include claims about the continued growth of the block and logical and other necessary truths. To the extent to which the future is open, it can only be a matter of contingent claims.

A popular way of modeling the alleged openness of the future is in terms of a branching time series. Suppose, again, that the present time is t_0. To say that the future is open is to say that various futures are possible at t_0 and that it has not yet been settled which of these possible futures will be realized. We can picture these various possibilities as a tree whose trunk is the past and whose branches represent the various possible futures.[3]

In such a branching time series, there are two main ways of interpreting the future tense operator 'F' ("it will be the case that"). Let a *possible future* be a maximal linear subset of times later than the present time t_0. In other words, a possible future is a way of drawing a straight line from t_0 through the entire branching future. Then one option for interpreting the future tense operator is to say that '$t_0 \mid F\varphi$' is true if and only if φ is true at a time in *some* such possible future. This "some" reading yields a standard tense logic for the future tense operator, but it has the peculiar consequence that all consistent predictions would be true. Since the branches at t_0 are supposed to include all possible futures, '$t_0 \mid F\varphi$' is true unless φ is impossible.

The second option is to say that '$t_0 \mid F\varphi$' is true if and only if φ is true at some time in *all* possible futures. On this "all" reading, '$F\varphi$' is only true at t_0 if φ is necessary. Neither '$t_0 \mid F\varphi$' nor '$t_0 \mid F\neg\varphi$' would be true if φ is contingent because φ would be false throughout some future histories, and true throughout others. To advocates of an open future, this will look attractive: the only part of the future that is settled at t_0 is what is necessary.

The tricky part is to spell out the rest of the "all" reading, which requires substantial changes to our logic. Since '$\varphi \vee \neg\varphi$' ("either φ or not φ") is a truth of standard logic, and thus true in all possible futures, '$t_0 \mid F(\varphi \vee \neg\varphi)$' should also be true. Yet if φ is contingent then '$t_0 \mid F\varphi$' and '$t_0 \mid F\neg\varphi$' are both false on the "all" reading of the future tense operator. To get a coherent account of the open future, we thus have to relinquish some part of standard logic, and there are number of different, and mutually incompatible ways of doing this. Popular proposal include the postulation of truth-value gaps (Thomason 1970), the adoption of a three-valued logic (Łukasiewicz 1970; Prior 1953; Bourne 2004), and the denial of truth-functionality (Tooley 1997).

[3] It is not at all clear what notion of possibility we need to employ for this purpose, but let me pass over this here; see Torre (2011) for more details.

These are complications, but they are complications that growing blockers should welcome. After all, their intention is to use a branching time-series to model the openness of the future without taking the picture seriously. They want to be anti-realists about a single future, not realists about multiple ones. The need to revise our logic about the future is precisely what marks our account of the future as *anti-realist* in the sense promoted by Michael Dummett (1991).

The problem is that none of this helps the *glowing* block view. Irrespective of our account of the truth conditions of future tense claims, our tense logic will always validate what Alex Malpass (2009) calls the *Future–Past Principle*:

$$\varphi \rightarrow FP\varphi \tag{8.5}$$

The model of time that we are considering postulates an asymmetry between past and future that is due to the fact that the time series only branches in the future direction. Because time does not branch backwards, there is no ambiguity in interpreting the past tense operator 'P' in this case. If φ is true now then we will always see φ in our past when we look back from any future time, which is what (8.5) asserts.

The Future–Past Principle formalizes the claim that the present will be the past. Since it postulates the existence of a future time at which 'Pφ' is true, and thus rules out the possibility that the present time is the last one, it is perhaps not entirely trivial. But since it is one way in which one can spell out the claim that the block will continue to grow, it is a principle that any growing blocker must accept as one of the necessary truths about the future, rather than as one of the contingent claims that are not yet settled.

If the Future–Past Principle is true then the glowing block view is false. If I am conscious now (and I am) then I will have been conscious, and it will be false that only beings present then have ever been conscious. The growing blocker might deny that there are future facts about consciousness, but what is at issue here are facts about the *present*. It follows from my being conscious now that there will be a time at which the past is not dead. No matter what future branch we end up in, when we look back in time we shall always see my present consciousness in our past. Being left behind by the flow of time does not turn one into a zombie—unless one is a zombie already.

It is worth noting that the growing block view can fail to vindicate the *Past–Future* Principle:

$$\varphi \rightarrow PF\varphi \tag{8.6}$$

Suppose we adopt the "all" reading of the future tense operator, and suppose further that φ is a contingent claim that is true now. Then it is not the case that φ was going to be true because at any earlier time there is at least one future-directed branch in which the contingent claim φ is false. So even though φ is true now, it was not going to be true, which conflicts with the Past–Future Principle.

Some philosophers find this result unpalatable. If we are now in the midst of a sea battle, they ask, how could it not be true that there was going to be a sea battle? Storrs McCall (1994: 3) tries to avoid this difficulty by postulating *branch attrition.* As time goes by, he claims, possible futures that were not realized get cut off from the tree of time. This ensures that the past always looks like a straight line, and thus validates the Past–Future Principle as well.

This is not the place to decide whether this is right, and whether the Past–Future Principle is in fact true. What is interesting for our purposes is that we can reject the Future–Past Principle if we are prepared to allow the time series to branch into the *past.* Ned Markosian (1995) argues that the only coherent sense to be made of the openness of the future is in terms of non-deterministic laws of nature. In this case, we could say that a future is possible if it is not ruled out by the present state of the world and the laws. But most laws of nature are time-reversal invariant. If this remains true for indeterministic laws then the past is bound to be as open as the future. There would be more than one possible past that is compatible with the present state of the world and the laws of nature. This would allow us to reject the Future–Past Principle, by adopting an "all" reading of the past tense operator '*P*'. If φ is a contingent claim that is true now then it does not follow that φ will have been true. At future times there will be past-directed branches in which φ is false, which means that (on an "all" reading of the past tense operator) '$P\varphi$' is false at future times. Turning this result around, this means that the Future–Past Principle encodes the *fixity of the past*, which is a central tenet of the growing block view. Yet the Future–Past Principle is precisely what rules out the *glowing* block view.

8.4

This paper argued that there is no interesting connection between consciousness and presentness. Whatever feature of consciousness we might focus on, it will always be the case that this feature can be instantiated at more than one time, and thus fails to single out a unique time as a privileged present. Moreover, once a time has acquired whatever feature is supposed to make it present, the Future–Past Principle prohibits it from losing it again in the future, and that is incompatible with the truism that different times are successively present. This does not show that human time is of no interest in the philosophy of time. In particular, it does not settle the question whether the phenomenon of the specious present tells us something interesting about the nature of time. But it does provide some support for the view that physical time is fundamental, that consciousness is just another physical process, and that the present is just another time.

References

Baker, Lynne Rudder. 2010. Temporal reality. In *Time and identity*, ed. Joseph Campbell, Michael O'Rourke, and Harry Silverstein, 27–47. Cambridge, Mass.: The MIT Press.

Bourne, Craig. 2004. Future contingents, non-contradiction, and the law of excluded middle muddle. *Analysis* 64: 122–128.

Buhusi, Catalin, and Warren Meck. 2005. What makes us tick? Functional and neural mechanisms of interval timing. *Nature Reviews Neuroscience* 6: 755–765.

Canales, Jimena. 2009. *A tenth of a second: A history*. Chicago: Chicago University Press.

Dainton, Barry. 2008. The experience of time and change. *Philosophy Compass* 3: 619–638.

Dainton, Barry. 2011. Time, passage, and immediate experience. In *The Oxford handbook of philosophy of time*, ed. Craig Callender, 381–418. New York: Oxford University Press.

Dolev, Yuval. 2007. *Time and realism: Metaphysical and antimetaphysical perspectives*. Cambridge, Mass.: The MIT Press.

Dolev, Yuval. 2012. Perceiving transience. In *The future of the philosophy of time*, ed. A. Bardon, 56–72. New York: Routledge.

Dummett, Michael. 1991. *The logical basis of metaphysics*. Cambridge, Mass.: Harvard University Press.

Dummett, Michael. 2000. Is time a continuum of instants? *Philosophy* 75: 497–515.

Dummett, Michael. 2005. Hume's atomism about events: A response to Ulrich Meyer. *Philosophy* 80: 141–144.

Forrest, Peter. 2004. The real but dead past: A reply to Braddon-Mitchell. *Analysis* 64: 358–362.

Forrest, Peter. 2006. Uniform grounding of truth and the growing block theory: A reply to Heathwood. *Analysis* 66: 161–163.

Grünbaum, Adolf. 1953. Whitehead's method of extensive abstraction. *British Journal for the Philosophy of Science* 4: 215–226.

Heathwood, Chris. 2005. The real price of the dead past: A reply to Forrest and to Braddon-Mitchell. *Analysis* 65: 249–251.

Hoerl, Christoph. 2009. Time and tense in perceptual experience. *Philosophers' Imprint* 9(12): 1–18.

James, William. 1890. *The principles of psychology*. New York: Holt.

Łukasiewicz, Jan. 1970. On three-valued logic. In *Selected works*, ed. L. Borkowski, 87–88. Amsterdam: North-Holland.

Malpass, Alex. 2009. Future tense in branching time. Unpublished manuscript.

Markosian, Ned. 1995. The open past. *Philosophical Studies* 79: 95–105.

Mauk, Michael, and Dean Buonomano. 2004. The neural basis of temporal processing. *Annual Review of Neuroscience* 27: 307–340.

McCall, Storrs. 1994. *A model of the universe: Space-time, probability, and decision*. Oxford: Clarendon Press.

Meyer, Ulrich. 2005. Dummett on the time-continuum. *Philosophy* 80: 135–140.

Myro, George. 1986a. Existence and time. In *The philosophical grounds of rationality*, ed. Richard Grandy and Richard Warner, 383–409. Oxford: Clarendon Press.

Myro, George. 1986b. Time and essence. In *Studies in essentialism*, Midwest studies in philosophy, vol. 11, ed. Peter French, Theodore Uehling, and Howard Wettstein, 331–341. Minneapolis: University of Minnesota Press.

Paul, L.A. 2010. Temporal experience. *Journal of Philosophy* 107: 333–359.

Prior, Arthur. 1953. Three-valued logic and future contingents. *Philosophical Quarterly* 3: 317–326.

Prosser, Simon. 2012. Why does time seem to pass? *Philosophy and Phenomenological Research* 85: 92–116.

Russell, Bertrand. 1915a. Our knowledge of the external world as a field for scientific method in philosophy. Chicago: Open Court.

Russell, Bertrand. 1915b. On the experience of time. *The Monist* 25: 212–233.

Thomason, Richmond. 1970. Indeterminist time and truth-value gaps. *Theoria* 36: 264–281.
Tooley, Michael. 1997. *Time, tense, and causation*. Oxford: Clarendon Press.
Torre, Stephan. 2011. The open future. *Philosophy Compass* 6: 360–373.
Whitehead, Alfred North. 1919. Time, space, and material. *Aristotelian Society Supplement* 2: 44–57.

Chapter 9
The Arrow of Time

Meir Hemmo and Orly Shenker

Abstract We argue that if everything there is in the world is physical, then time has an objective direction. If the fundamental equations of motion are time reversal invariant we show that the direction of time cannot be explained by anything else in physics (e.g. the direction of processes in time) and therefore must be added to physics. We further argue that the direction of time gives rise to both the thermodynamic and the psychological arrows of time (whenever they exist), and that it is necessary in order to construct a meaningful Past Hypothesis in statistical mechanics.

Keywords Experience of the direction of time • Direction of time • Entropy • Past Hypothesis • Time-reversal invariance • Second Law of Thermodynamics

9.1 Introduction

Two facts about our experience are uncontroversial. (i) We experience a direction of time: there is a clear distinction between past and future in our experience. (ii) We experience a direction of processes relative to this direction of time. One type of such processes is described by the Second Law of Thermodynamics according to which the entropy of thermodynamic systems cannot decrease over time. Usually the Second Law of Thermodynamics is not taken to be a fundamental law of physics, and one seeks an account of it in terms of more fundamental theories such as classical or quantum statistical mechanics. The problem is that the fundamental laws of physics are invariant under the reversal of the direction of time (TRI), and therefore it is not clear how could such theories underwrite the thermodynamic direction of

M. Hemmo (✉)
Department of Philosophy, University of Haifa, Haifa 31905, Israel
e-mail: meir@research.haifa.ac.il

O. Shenker
Program in the History and Philosophy of Science, The Hebrew University of Jerusalem, Mount Scopus, Jerusalem 91905, Israel

Y. Dolev, M. Roubach (eds.), *Cosmological and Psychological Time*, Boston Studies in the Philosophy and History of Science 285,
DOI 10.1007/978-3-319-22590-6_9

time. Moreover, since as we said it is a fact of our experience that time has a direction, it is not clear how this fact can be underwritten by fundamental laws that are TRI, and therefore don't distinguish between past and future.

The problem becomes even more acute if one assumes physicalism about our experience (and about our mind in general). The very fact that we have an experience of a temporal arrow suggests one of the following two options. (i) If the mental is physical, then the fundamental laws can't be TRI, and so physics has to be changed or supplemented with an arrow of time; (ii) The mental is *not* physical. Our point in this paper is that in currrent physics there seems to be no other way to explain our experience of the direction of time. In a physicalist framework there has to be something physical or some physical fact that will give rise to this experience. This fact can either be in our environment or within the brain or both – but in all cases it has to be in the physics of things, and our experience of the direction of time may then supervene on this physical fact. If there is no such fact, then assuming physicalism there has to be an objective physical arrow of time in addition to all the other facts of physics.

In Sect. 9.2 we explain what time reversal invariance means in classical mechanics focusing on kinematics only. In Sect. 9.3 we argue that theories in which the laws of motion are time reversal invariant cannot account for the physical facts, which underwrite our experience of the direction of time. In this sense such theories are incomplete. We discuss the direction of thermodynamic processes in time such as the process in which entropy increases in Sect. 9.4. We then show that the Second Law and the Past Hypothesis in statistical mechanics cannot yield arrows of time since they assume an arrow of time. Our conclusion is that an arrow of time has to be added to physics, in one way or another. In this sense current physics is not complete, and its lacuna is in a very central and conspicuous place in the empirical data.

We don't offer the details of where and how exactly to add the missing arrow. We only point out one requirement that seems to us necessary: if the theories are taken to describe time slices of some sort, then the arrow must be part of each and every time slice since such an arrow cannot be derived from the order or overall structure of the series of slices.

9.2 Time Reversal Invariance

Perhaps the simplest way of demonstrating what is meant by time-reversal invariance is by looking at the kinematics of classical mechanics. We shall focus on an example of the most elementary mechanical evolution, namely, the free evolution of a single particle that is located in point A in space at time α and at point B in space at time β. In intermediate times the particle is located in points in space and time that satisfy a relation of betweeness that is described by the classical equations of motion. We call the state of the world at each such time *time-slice*. As pointed out by Malament (2004) (see also our 2012, Chap. 4) the description of the evolution so far is ambiguous in the sense that one can say either that the particle moves from A to B or from B to A. Resolving this ambiguity requires that we fix a direction of time (or temporal orientation). We shall address the question how this should be done at the end of this section. For now, let us suppose that α is earlier than β. Given this direction we can now say that the velocity of the particle is in the spatial direction from A to B. If we were to set the direction of time such that β preceded α, then the velocity of the particle would be from B to A. We shall consider the constraints (if any) on the choice of the direction of time later. For now let us suppose that the direction is from α to β. Call this evolution the "forward" evolution. Let us now consider some *temporal* relations that are associated with this description: velocity reversal, retrodiction and time reversal.[1]

By *velocity reversal* of the "forward" evolution one means that the spatial direction of the velocity is reversed, for example if a particle has a velocity that takes it from position A to position B, then a reversal of its velocity means that the particle will go from position B to position A given the *same* direction of time, namely the direction from α to β. That is, in velocity reversal we reorder the set of the time slices in the figure such that the time slice in which the particle is in position B is set to be first (i.e. its time coordinate is now α) and the time slice in which the particle is in position A is set to be last (i.e. its time coordinate is now β). The spatial direction of the velocity of the particle is thus reversed (that is, it is now in the direction from B to A) relative to the same direction of time we started with.

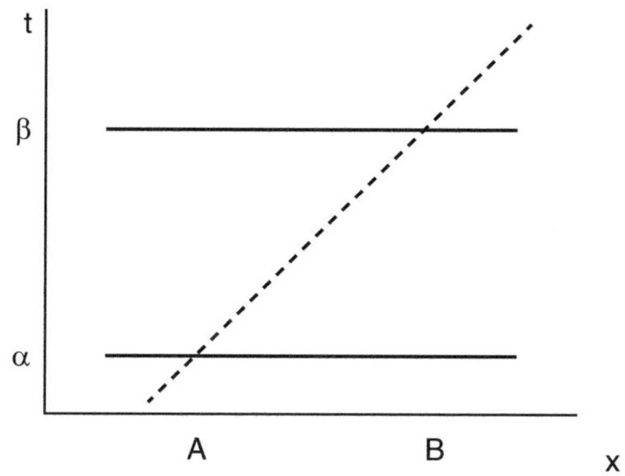

By *retrodiction* of the "forward" evolution one means that given the temporal order of the time slices that we started with (namely from α to β) and given the time slices exactly as they appear in the "forward" evolution, and given the last time slice (at β) one calculates the preceding time slices. That is, the time slices and the direction of time remain exactly the same as in the "forward" evolution, but the input of the calculation is the end time slice of the forward evolution and the output is the

[1] We focus here on velocity reversal in kinematics, and we do not address the significance of the symmetry of other theories under velocity reversal. In particular we do not discuss the fact that in order for a particle to return from *B* to *A*, if its evolution from *A* to *B* was affected by a magnetic field, one needs to reverse not only the velocity but also the field. See discussions on this topic in for example Albert (2000, Chap. 1), Earman (2002), Malament (2004) and Arntzenius and Greaves (2009).

initial time slice. Two things are important to note here: first, the direction of time remains the same as in the "forward" evolution. Second, and more importantly, retrodiction is a re-description (or inference) of the "forward" evolution. It does not correspond to a new physical process distinct from the "forward" evolution, since the calculation goes from a time slice in which the particle is in position B to a time slice in which the particle is in position A, while the velocity is in the spatial direction from A to B.

By *time reversal* one means the following. First of all we set the direction of time to be from β to α. That is β now is considered earlier in time than α. Therefore, the time reverse of the "forward" evolution is an evolution in which the particle starts in position B at time β and evolves dynamically (not by retrodiction) by the equations of motion to position A at time α.[2] Now: we say that a theory is time reversal invariant if the laws of the theory remain the same under time reversal.

We now comeback to the question of how the direction of time should be fixed. Let us distinguish between theories that are time reversal invariant and theories that are not. In time reversal invariant theories there is *no physical fact* that prefers the direction of time, that is the direction from α to β or the other way around. While the temporal relation of betweeness is determined by the equations of motion, the direction of time is a matter of *convention*. That is, given everything there is in classical mechanics for example, there is no fact of the matter as to whether the particle evolves from A to B in the forward direction of time or from B to A in the backward direction. If one accepts classical mechanics (or any other time reversal invariant theory) as a complete theory of the world, then one is committed to the claim that there is no fact of the matter as to the direction of time. Is this a reasonable conclusion given what we know about the world?

9.3 The Direction of Time

In short we think that the above conclusion is unreasonable given what we know about the world. The reason is that we know that there are clear and distinct facts in the world that are completely missing in the above account. What are these facts? Consider an *observation* of the evolution of a particle that takes place between α and β. In every occasion of such an evolution *we* can say immediately and directly whether α is in the past or in the future. That is, we can say in every occasion whether for example the particle was in the past in position B or in position A.[3] This

---

[2] One may wonder how to express precisely this reversal, in particular whether the double reversal of sings result in a *velocity* with the same sign as in the original forward evolution. The answer is that this depends on some conventions, (see our 2012, Chap. 4).

[3] One way in which we come to have this knowledge is by memory. But by this we do not mean that we define the events that we remember to be in the past: this is characteristic but not analytic. It is physically possible given the laws of classical mechanics for example to remember events that occur in the future. See our (2012, Chap. 10), and compare Hawking (1985, 1988).

is what we meant in our opening sentence when we said that we have an experience of the direction of time. The facts that we are talking about are *psychological*. And we take these psychological facts as a starting point in this discussion. It is a matter of empirical observation that this is the sort of experience that we have. In fact, it is one of the most salient features of our experience that whatever it is we have an experience of is temporally ordered. This feature of our experience seems to us so obvious and fundamental that we have no idea what it would take to argue for it. If you deny this you are bound to say that our sense of time (whatever it is) consists of only the relation of betweeness. And what we assume here is that this is not true, as a matter of primitive fact about our psychology.

So given that our experience is temporally directed, what feature of the world underwrites (or gives rise to) this phenomenon? Let us suppose that whatever this feature is, it should be accounted for in physics.[4] Imagine now two worlds, one in which the direction of time is "forward" and the other in which the direction of time is "backward". Suppose that the world (including us) is situated in an intermediate time slice between α and β, call it γ. And we want to know which time slice is earlier, α or β. The equations of motion will not distinguish between these two possibilities since they are time reversal invariant. Now suppose (as we do) that we sense that, for example, α precedes β. This is what we mean by assuming that there are psychological facts about our experience of the direction of time. Assuming that everything there is in the world is physical (and can be accounted for by physics) these psychological facts should be underwritten by physical facts. Since our sensation is part of γ this means that the temporal relation between α and β should be somehow encoded in γ. What can this coding consist of? It cannot be the positions in γ since the positions are the same as in the time-reversed world. All that is left in γ are the velocities, but these velocities assume a direction of time, if they correspond to unambiguous facts (see above). And so the encoding in γ of the direction of time must be an *additional* fact about γ over and above the equations of motion and the positions.[5] Note that it is a triviality that not all the physical facts in γ are expressed by the equations of motion. For example, the positions in γ are facts that one must add in addition to the equations of motion. More generally, the notion of initial conditions means that one adds positions as well as a direction of time in order to solve the equations of motion. Laws that are invariant under the reversal of time are simply laws that are insensitive to this additional feature of the time slices. But never mind the laws: *we* are sensitive to this feature.

If, however, the laws of the underlying theory are not time reversal invariant (as in some quantum mechanical theories involving for example the collapse of the quantum state), one can say the following. In every evolution we still have an experience of the direction of time (just like before in the time reversal invariant theories),

---

[4] Of course this may not be true, but here we wish to assume that everything in the world is physical. Also there are various views about how precisely this physicalist idea should be cashed out, but our argument does not depend on how these differences will be resolved.

[5] We assume here that there is a well-defined notion of an instantaneous state that includes velocity. For various approaches to this question, see Arntzenius (2000).

but in such theories one can place the origin of this experience in the differences of the evolutions relative to the two directions of time. Since in this case the laws are sensitive to these differences, one may say that also our experience is sensitive to these differences.

9.4 The Past Hypothesis

It is sometimes argued that our experience of the direction of time can be accounted for on the basis of the Second Law of Thermodynamics. We will now explain very briefly why this view is mistaken.

The statistical mechanical account of the Second Law of Thermodynamics consists of two parts, which differ markedly in their contents as well as in their argumentative structure. The first part concerns the thermodynamic prediction of the increase of entropy towards the future, and the second part concerns the thermodynamic retrodiction that entropy has been lower in the past. The first part – the predictive part – is often explained as follows.

Consider the paradigmatic case of a gas confined by a partition to the left half of an isolated container. We know from experience that if we remove the partition the gas will fill the entire volume of the container after a very short time and will remain in this state indefinitely. This is an example of the principle of the approach to equilibrium and of the Second Law of Thermodynamics, according to which, respectively, isolated systems approach equilibrium and remain in this state indefinitely, and the entropy of the final equilibrium state of these systems cannot be lower than the entropy at their initial equilibrium state (see Brown and Uffink 2001 for this distinction). The evolution described by these principles is asymmetric in time: processes such as the expansion of the gas never happen in the reversed order. Statistical mechanics attempts to explain this time-asymmetric behavior on the basis of the laws of classical (or quantum) mechanics governing the behavior of the particles that make up the gas. It is well-known today that this aim cannot be achieved on the basis of classical mechanics alone, and the best one may hope for is to show that – given the laws of mechanics – the behavior described by the principles of thermodynamics is highly *likely*, and its probability is so high that we ought not to expect to see anything else.[6]

The second, retrodictive, part, as is well known suffers from a problem known as the *minimum entropy* (or *parity of reasoning*) problem. Let us explain the problem and describe how it is solved by the, so-called, Past Hypothesis. The probabilistic validity of the principles of thermodynamics is meant to supervene on the underlying dynamics, and the nature of *retrodiction* in mechanics (as described above)[7] entails that any proof in mechanics that entropy has a certain probability to increase

[6] The extent to which this probabilistic claim can be justified is addressed in Hemmo and Shenker (2012), especially Chaps. 7, 8, and 13.

[7] Not of time reversal; as we have shown these are physically distinct notions despite their mathematical equivalence.

towards the *future* is also a proof that entropy has the same probability to increase towards the *past*. That is, the entropy of the present state is invariably minimal, for any present state; hence the names 'the minimum problem' and the 'parity of reasoning problem'. However, the increase of entropy towards the past is inconsistent with the empirical data: our memories and records tell us that entropy was lower in the past. For example, once the gas in the above example has reached its final state, we remember that it was prepared in a macrostate of entropy *lower* than the entropy of the final state, and that it has spontaneously expanded.

This problem is solved in statistical mechanics by the so-called Past-Hypothesis.[8] Here is Richard Feynman's way of expressing it:

> ...I think it necessary to add to the physical laws the hypothesis that in the past the universe was more ordered, in the technical sense, than it is today.[9] (In Feynman's terms, "more ordered states" are states of lower entropy)

That is, Feynman's proposal is to add to the theory the assumption of the low entropy in the past despite the fact that it is highly *un*likely (by the parity of reasoning argument; see above) that the low entropy in the past actually obtained. Forget for a moment the *ad-hoc* flavor of this idea; this flavor can be dealt with, in a way that we will not discuss here.[10] The problem we would like to focus on is this: *where* in time – in what *direction* of time – lies the past that Feynman is talking about? *Which side of the present is past?* The tautological answer – that the past is in the past relative to the present – won't do here, since although we can point at the past by using our inner perception, we cannot point at it in terms of our physical theories (if they are time reversal invariant). And our question was, what aspect of physics gives rise to our experience. And since within the Past Hypothesis – which is part of the statistical mechanical account of the Second Law of Thermodynamics – there is already a reference to a past, it is clear that we cannot turn to the Second Law in order to look for the direction of time; this would be a viciously circular move. And so, in order to make sense of the Past Hypothesis, there must be a way of *breaking the symmetry* between past and future in classical mechanics. This symmetry breaking can then be used to express the asymmetry in time of our experience. And the crucial point here is that there is no need to add a special *ad-hoc* symmetry-breaking feature to the theory, since we already have such a feature: the symmetry breaking feature of mechanics is the arrow of time in each of the elementary time slices. As we saw, adding such an arrow to the elementary description of a mechanical microstate is necessary in order to account for our experience of the direction of time. This arrow serves to accommodate the Past Hypothesis in statistical mechanics and solve the problem of breaking the asymmetry between past and future.

In sum, by introducing the Past Hypothesis one is committed to two statements: one is that in the past the entropy was lower. It is a contingent fact about our world that is supported by our memories of the past and need be added over and above the

---

[8] See Feynman (1967), Albert (2000, Chap. 4), Callender (2003), Earman (2006).

[9] Feynman (1967), p. 116.

[10] See Hemmo and Shenker (2012, Chap. 10).

laws of mechanics and the fact that we perceive time as having a preferred direction.[11] The other, which is presupposed by the first statement, is that there is a direction of time, namely there is an objective matter of fact about which direction of time is the past and which is the future. This is necessary if we wish to have a physical underpinning of our experience of the direction of time even if all the other laws of our theory are invariant under a reversal of the direction of time.

Acknowledgements This research has been supported by the Israel Science Foundation, grant number 713/10, and by the German-Israel Foundation, grant number 1054/09.

References

Albert, David. 2000. *Time and chance*. Cambridge: Harvard University Press.

Arntzenius, Frank. 2000. Are there really instantaneous velocities? *The Monist* 83: 187–209.

Arntzenius, Frank, and Hilary Greaves. 2009. Time reversal in classical electromagnetism. *British Journal for the Philosophy of Science* 60(3): 557–584.

Brown, Harvey, and Jos Uffink. 2001. The origins of time-asymmetry in thermodynamics: The minus first law. *Studies in History and Philosophy of Modern Physics* 32(4): 525–538.

Callender, Craig. 2003. Is there a puzzle about the law entropy past? In *Contemporary debates in the philosophy of science*, ed. C. Hitchcock. Oxford: Blackwell.

Earman, John. 2002. What time-reversal invariance is and why it matters. *International Studies in the Philosophy of Science* 16(3): 245–264.

Earman, John. 2006. The Past Hypothesis: not even false. *Studies in History and Philosophy of Modern Physics* 37: 399–430.

Feynman, Richard. 1967. *The character of physical law*. Cambridge, MA: MIT Press.

Hawking, Steven. 1985. The arrow of time in cosmology. *Physical Review D* 32(10): 2489–2495.

Hawking, Steven. 1988. *A brief history of time*. London: Bentam Press.

Hemmo, Meir, and Orly Shenker. 2012. *The road to Maxwell's Demon*. Cambridge: Cambridge University Press.

Malament, David. 2004. On the time reversal invariance of classical electromagnetic theory. *Studies in History and Philosophy of Modern Physics* 35B(2): 295–315.

Price, Huw. 1996. *Time's arrow and Archimedes' point: New directions for the physics of time*. New York: Oxford University Press.

[11] The fact that we have memories of the past answers Price's (1996) charge of a double standard since the symmetry between the past and the future is broken. Nevertheless it is a contingent fact that we remember the past rather than the future, but we shall not go into this issue here (see footnote 3 and Hemmo and Shenker 2012, Chap. 10).

Part III
Temporality and Phenomenology

Chapter 10
Heidegger's Primordial Temporality and Other Notions of Time

Michael Roubach

Abstract This paper examines the possible contribution of Heidegger's notion of "primordial temporality" to elucidation of the general concept of time. It begins by surveying some possible motivations for adopting this sort of notion, with special emphasis on motivations arising from awareness of the problem of consciousness of time. It then addresses Heidegger's claim that primordial temporality is the basic concept of time, and that other concepts of time, and specifically, what Heidegger calls the "ordinary" (*vulgäre*) notion of time, presuppose it. Some scholars contend that Heidegger does not present an adequate justification for this claim. Invoking affinities between Heidegger's position and Brouwer's intuitionism, the second half of the paper puts forward an argument for Heidegger's claim that the ordinary notion of time presupposes primordial temporality.

Keywords Heidegger • Primordial temporality • Continuum • Brouwer

In *Being and Time*, Heidegger develops a notion of time he calls "primordial temporality" (*ursprüngliche Zeitlichkeit*). His purpose in doing so is primarily to provide a framework for the meaning of Being. This paper will consider the possibility that the notion of primordial temporality plays a crucial role, not only within the Heideggerian approach to time, but also in the elucidation and exposition of the concept of time in general. After briefly explaining Heidegger's notion of primordial temporality, I will survey some possible motivations for adopting this sort of notion, with special emphasis on motivations arising from awareness of the problem of consciousness of time.

The paper's major focus will be Heidegger's claim that primordial temporality is the basic concept of time, and that other concepts of time, and specifically, what Heidegger calls the "ordinary" (*vulgäre*) notion of time, presuppose it. Ordinary time is characterized as "an endless, irreversible sequence of 'nows' which passes

M. Roubach (✉)
Department of Philosophy, Hebrew University of Jerusalem, Jerusalem 91905, Israel
e-mail: roubach@mail.huji.ac.il

Y. Dolev, M. Roubach (eds.), *Cosmological and Psychological Time*, Boston Studies in the Philosophy and History of Science 285,
DOI 10.1007/978-3-319-22590-6_10

away" (SZ 426). This sequence is arranged in accordance with earlier than/later than relations between 'now's, and, inasmuch as there is no end to the 'now's, it is infinite. Both Paul Ricoeur (1988, 87–88) and William Blattner (1999, 181–184) maintained that Heidegger did not provide a satisfactory argument for his claim about the primacy of primordial temporality. I will show, however, that on the strength of certain assumptions that can be reasonably ascribed to Heidegger, an argument for the primacy thesis can be constructed, though the argument as I present it is not explicitly put forward in Heidegger's writings. The argument is particularly edifying in that it relates Heidegger's discussions of the various notions of time to the more general issue of the arithmetization of continuous time.

Primordial temporality is the temporality of Dasein, the entity who is preoccupied with the question of Being. It is not the time-span of Dasein's existence, but rather, it is Dasein's relationship with time, that is, the manner in which Dasein understands its temporality.

Various characterizations of this primordial temporality are given in *Being and Time*, but they can be classed under two main rubrics: futurity and finitude.

In primordial temporality, the future is the basic temporal 'dimension': it is in terms of the future that the present and past are understood. However, the future is not "a 'now' which has *not yet* become 'actual,'" as it is usually understood, but is characterized as anticipation: "Anticipation makes Dasein *authentically* futural" (SZ 325).[1] The meaning of "future" is "[coming] towards-oneself," it is the move toward oneself that is made by evincing what Heidegger calls "resoluteness" (SZ 329), namely, by Dasein's authentic self-disclosure (SZ 296). This future is characterized as "towards-oneself" because it separates Dasein from other entities by turning Dasein toward its own being, toward its death, a separation that requires resoluteness. Heidegger's basic characterization of the future is, therefore, anticipatory resoluteness (SZ 326).[2] But if, in primordial temporality, the future is basic, what is the meaning of the past? The meaning of the past in primordial temporality is "having-been-ness" (*Gewesenheit*). The past is understood through the future, because it is revealed through Dasein's relation to its own death, its future past-ness. As "having-been-ness," the past situates Dasein and enables Dasein to encounter that which is given in its environment (SZ 326). Dasein's temporality as a whole, Heidegger tells us, is "the unity of a future which makes present in the process of having been" (SZ 326).

The second main feature of primordial temporality is finitude. This finitude is not characterized in terms of finite duration, but rather in terms of the limitedness of Dasein's existence. Dasein's time is finite in the sense that its future, its coming toward itself, is made possible by its "Being-towards-death" (SZ 329–330). Its future is thus characterized by a possibility (Dasein's death) that cannot be outstripped. This possibility cannot be realized by Dasein itself (SZ 330). Heidegger calls death, as this unique kind of unrealizable possibility, "extreme possibility."

[1] References to Heidegger 1962 will cite the page number of the original German edition.

[2] Heidegger upholds the token-reflexivity of time without basing it primarily on the 'now.'

Death is certain, but also indefinite, in the sense that we do not know when it will arrive.

Rephrasing the notion of primordial temporality, we can say that it is one's understanding of her own life as it "*stretches along between* birth and death" (SZ 373), but without assuming the usual interpretation of this interval as an orderly succession of points. The limit of life is death. Death can come at any moment and impose its irrevocable determination, but death itself is not an integral part of life. Death is, rather, the horizon of all our possibilities.

Why does Heidegger develop this notion of primordial temporality? Primordial temporality is intended to provide a transcendental horizon for the question of the meaning of Being. I cannot go into all the details of Heidegger's understanding of this question, which is a central—perhaps *the* central—theme of Heidegger's thought, but we must take note of an aspect of the question of Being that is relevant to primordial temporality. The question of Being, Heidegger tells us, is inextricably linked to Dasein's understanding of that question. A key component of this understanding arises from Dasein's understanding of death, through which Dasein distinguishes its existence from non-existence. Heidegger maintains that this specific understanding of Dasein's own existence does not presuppose any previous understanding of the meaning of Being in general (SZ 241–242).

I now turn to the question of whether the notion of primordial temporality is interesting, not merely from the perspective of Heidegger's thought, but also from the perspective of the philosophy of time in general. To answer this question, I will explore the relationship between, on the one hand, primordial temporality, and on the other, consciousness of time and ordinary time, without necessarily confining myself to the specific ways Heidegger himself characterizes the temporal concepts of ordinary time and world time (*Weltzeit*). The said relationship can be viewed in two opposed ways. It could be argued that primordial temporality represents a certain aspect of our understanding of time, an aspect related to the manner in which way one conceives and shapes one's own life, but not necessarily relevant to the more fundamental essence of the concept of time. For example, it might be claimed that primordial temporality describes some aspects of our consciousness of time, but not the basic experience of perceiving change. This line of argument challenges Heidegger's primacy thesis. Alternatively, we might seek to support that thesis by examining whether primordial temporality can be shown to be pertinent to the consciousness of time. If so, this will indicate that it does, indeed, play a foundational role in our experience of time.

10.1 Time Consciousness and Primordial Temporality

Heidegger's discussions of the concept of time do not address consciousness of time directly. But we can gain some knowledge of Heidegger's outlook on this issue by considering his attitude to Husserl's position on consciousness of time. Around the time *Being and Time* was published (1927), Heidegger edited Husserl's lectures on internal time consciousness. In *Being and Time* there is no explicit discussion of either Husserl's views on time consciousness, or issues pertaining to time consciousness more generally, but some of the concepts Heidegger uses in his discussions of time seem to be related to those used by Husserl in discussing time consciousness. For example, in describing the existential-ontological interpretation of Aristotle's notion of time, Heidegger adduces the notions of retaining (*Behalten*) and awaiting (*Gewärtigen*) (SZ 421), which seem to have an affinity with Husserl's notions of retention and protention.

In his 1928 lectures, however, Heidegger mentions Husserl's lectures explicitly (Heidegger 1984, 203–204). Heidegger contends that Husserl's treatment of time consciousness rests on an assumption about the nature of time, namely, that it is composed of moments. This is an assumption Heidegger rejects:

> Expectancy (*Gewärtigen*), retention (*Behalten*), and making present (*Gegenwärtigen*) are not merely the way we grasp the then, the formerly, and the now, not merely modes of being conscious of them; they are rather the very origin of the then, the formerly, and the now. Expectancy is not a mode of being conscious of time but, in a primordial and genuine sense is time itself. (Heidegger 1984, 203)

Our understanding of time, Heidegger argues, does not presuppose any other temporal notion, and specifically, does not presuppose the ordinary notion of time. Moreover, according to Heidegger, time is inseparable from our understanding of time. Heidegger's discussions of time all invoke our experience and understanding of time, since, from the phenomenological perspective, a notion of time that transcends or is completely divorced from experience cannot be justified.

From Heidegger's point of view, the problem of time consciousness, as framed by Husserl, is premised on unacceptable assumptions about time. The problem of time consciousness is usually formulated as follows: if time consciousness is a fleeting grasp of a moment, then it seems that the change inherent in the notion of time is lost. On the other hand, if consciousness of time requires a temporal interval, then it is not clear how the *flow* of time can be grasped.

Can Heidegger's primordial temporality help provide a solution to this problem? Heidegger's position is that our consciousness of time is based primarily on the future, not the present. The present cannot serve as the foundation for time consciousness, since it presupposes the experience of time. In other words, consciousness of time is not centred on the 'now.'[3]

---

[3] Shaun Gallagher interprets Heidegger's primordial temporality as an alternative to Husserl's account of time consciousness, but his account differs from my own in focusing on the distinction between authentic and inauthentic experience and not on the dependence of ordinary time on primordial temporality. See Gallagher 1998, 115–120.

10.2 Primordial Temporality and the Ordinary Notion of Time

Heidegger's discussion of time consciousness rests on his claim that the ordinary conception of time as an endless sequence of 'now's is grounded in primordial temporality. Let us examine this view more closely, beginning with Heidegger's conception of the ordinary notion of time:

> This time is that which is *counted* and which shows itself when one follows the travelling pointer, counting and making present in such a way that this making-present temporalizes itself in an ecstatical unity with the retaining [*Behalten*] and awaiting [*Gewärtigen*] which are horizontally open according to the "earlier" and "later". (SZ 421)

In this sentence, Heidegger combines the Husserlian notions of retention and protention with the Aristotelian definition of time as the number of motion in relation to "earlier" and "later." According to Aristotle, the soul's 'contribution' to the notion of time is the numbering of 'now's. We can therefore say that consciousness of time as ordinarily conceived is constituted by the act of numbering, together with retention of past 'now's and anticipation of future 'now's. Furthermore, we can also conclude that, in Heidegger's opinion, the Husserlian model of time consciousness does not add anything substantive to the Aristotelian definition of time.

Heidegger maintains that the ordinary conception of time is dependent on primordial temporality. His principal argument for this position is that the ordinary notion of time is arrived at via Dasein's fleeing death by averting its gaze, by looking away from (*Wegsehen von*) its finitude. This looking away is a death-oriented relationship between Dasein and its future, and it is in this sense that the ordinary notion of time presupposes primordial temporality (SZ 424).

I will take another approach to explaining the relationship between ordinary time and primordial temporality—that is, the former's dependence on the latter—an approach centred on the time–number relation. This approach helps clarify Heidegger's contribution to the broader philosophical discourse on the concept of time.

Heidegger asserts that time and number are linked on two different levels: ordinary time and world time. World time is time as experienced and used in daily life. One feature of world time is datability. Datability is structured around a time's being "for" some activity: 07:00 is the time for waking up. We generally use clocks to situate ourselves relative to these different "for"s. On the one hand, datability is grounded in the "ecstatic" character of primordial temporality, i.e., the idea that temporality is always outside of itself (SZ 329); on the other, it is the basis for counting time as time is ordinarily conceived. In time as ordinarily conceived, the "for" is lost: it gets "levelled off" (SZ 422), and we are left with nothing but an enumeration of 'now's.

Invoking world time as a link mediating between primordial temporality and the ordinary notion of time explains how, given primordial temporality—which, we saw, is characterized by finitude—the notion of ordinary time, which is infinite, can be arrived at. But it does not explain why *infinite* time must be grounded in *finite* time.[4] That is, it does not explain why ordinary time *requires* finite time. This lacuna is at the core of Ricoeur's and Blattner's respective critiques of Heidegger's thesis about the primacy of primordial temporality.

It is, however, possible to provide an argument for the dependence of infinite ordinary time on finite primordial temporality, an argument that starts out by reflecting on the meaning of temporal infinity. The infinity of time is typically understood as limitlessness, as inexhaustibility—there is always more time, or to put it differently, there is no last 'now.' What does this understanding of temporal infinity imply? Does it mean that for any 'now' there is another, later, 'now'? If this is the meaning of infinite time, it would seem that the acknowledgement of temporal infinity requires an external, extra-temporal, vantage point. But is this the only option? Can time's infinity be understood only from an extra-temporal perspective? In conceptualizing time, is some variant of Plato's characterization of time as "the moving image of the eternal" unavoidable?

Heidegger presents his approach as an attempt to understand time without assuming any extra-temporal point of view. In "The Concept of Time," a lecture delivered in 1924, he contrasts the theologian, who understands time from the perspective of eternity, to the philosopher, who is "resolved to *understand time in terms of time*" (1992, 1–2). *Being and Time* develops this line of thought. In essence, Heidegger argues that the ordinary concept of time cannot, in itself, account for time's inexhaustibility. How can we grasp that the future is *always* open without assuming an extra-temporal point of view?

Heidegger's response to this challenge can, I contend, be found in the notion of primordial temporality. This reading of Heidegger's position is not expressly manifest in *Being and Time*, where he is trying to show how Dasein grasps infinite ordinary time by "looking away from the end of Being-in-the-World" (SZ 424). But it is in line with the approach Heidegger takes in *The Basic Problems of Phenomenology*, where he seeks to show how the ordinary notion of time arises from Dasein's temporality (*Zeitlichkeit*).[5] Heidegger claims that "the essential countedness of time is rooted in the ecstatic-horizonal constitution of temporality" (1988, 274). By this he means that time's infinity, its inexhaustible supply of 'now's, arises from Dasein's transcendence, Dasein's ex-sistence. This ex-sistence, however, arises from Dasein's temporality. Heidegger can thus argue that the ordinary notion of time is grounded in primordial temporality without having to assume an extra-temporal point of view.

Blattner finds this argument problematic, contending that Heidegger cannot account for the characterization of world time (and hence, ordinary time) as a

---

[4] Blattner (1999, 181–182) presents the problem as follows: world time's chains of "for"s are limited, i.e., finite. An orderly sequence of world time can be generated from these chains only if the infinite sequence of the ordinary notion of time is already presupposed.

[5] In *The Basic Problems of Phenomenology*, Heidegger distinguishes between temporality (*Zeitlichkeit*) and Temporality (*Temporalität*). This distinction points to two roles played by primordial temporality. The first of these notions (*Zeitlichkeit*) is limited to the grounding of Dasein's being, whereas the second (*Temporalität*) "is intended to indicate that temporality, in existential analytic, represents the horizon from which we understand being" (1988, 228). It is the latter notion that is pertinent to the dependency relationship I am trying to show here.

sequence. The connectedness of the different 'now's into one sequence cannot, he claims, be accounted for on the basis of primordial temporality alone. "This sequence of world-time times … is not imposed by any for-the-sake-of-which. It is only a result of having to fit several tasks into a given time sequence. But, of course, it is precisely the sequentiality of time that we are trying to explain" (1999, 181–182). Formulating this critique in non-Heideggerian terms, the transitivity of the instants in ordinary time cannot be accounted for on the basis of primordial temporality.

It seems to me that Blattner's interpretation of the relationship between the two modes of time in question may be too strong. Blattner requires every feature of ordinary time to be grounded in primordial temporality, whereas I would argue that a weaker sort of correspondence is sufficient to establish Heidegger's claim. Clearly, though, a close connection between the two modes of time must be demonstrated if we are to accept Heidegger's claim.

However, even if we accept Blattner's critique, I would like to suggest another argument for the primacy of primordial temporality, that is, another argument for ordinary time's dependence on primordial temporality. This argument invokes the fact that ordinary time is not only a sequence of 'now's, but also continuous. Heidegger himself did not put forward this continuity-based argument, but might well have endorsed it, since he was emphatic about the continuity of ordinary time: "The sequence of 'nows' is unbroken and has no gaps. No matter how 'far' we proceed in 'dividing up' the 'now,' it is always now" (SZ 423). The argument adduces a feature of Brouwer's intuitionism, specifically, Brouwer's notion of the continuum, which is articulated in terms of choice sequences. More specifically still, the argument adduces Carl Posy's interpretation of Brouwer's position, as well as Posy's notion of "forced indeterminacy," to interpret Heidegger's position.

My argument is based on certain affinities between both the problems addressed by Brouwer and Heidegger, and the means used to resolve them. These affinities include: rejection of the universal validity of the law of excluded middle; interpreting mathematical objects as temporal[6]; rejection of the view that the continuum is a set of points (Heidegger discusses this in several places in his early lectures, mentioning Cantor, Russell, and Weil; see, e.g. 1997, 80–81; 2007, 61–62)[7]; and rejection of an eternalist, extra-temporal perspective. In the introduction to *Being and Time*, Heidegger mentions the formalism–intuitionism controversy. Oscar Becker, who studied under Husserl and Heidegger, took note of the affinity between the views of Brouwer and those of Heidegger. In his *Mathematische Existenz*, Becker invokes Heideggerian notions to interpret Brouwer's intuitionism. He contends, for example, that Brouwer's distinction between sequences generated by a rule and choice sequences is meaningful only on the assumption of a finite perspective (Becker 1927, 230).

Brouwer's work on the intuitionistic continuum is from what is often called his mature period (from 1917 onward), during which he developed the concept of a

[6] For a discussion of this point, see Roubach 2008, 53–54.

[7] Heidegger was probably acquainted, through Oscar Becker's work on geometry, with contemporary discussions about the continuum problem; see Becker 1923.

choice sequence, and applied it to the continuum. A choice sequence is a sequence whose elements are freely (or randomly) chosen. For example, consider a number whose decimal expansion is 0.352866631….; it is generated by free choice of a digit at each stage. Choice sequences are potentially infinite. They provide an arithmetical form for the continuum, transforming it into a set of points. Inasmuch as they evolve over time, they are temporal entities. The notion of time that corresponds to such a choice sequence necessarily includes an open future. This is a point on which the affinity between the respective approaches of Brouwer and Heidegger is apparent, because for Heidegger, one of the key aspects of ordinary time is counting. Choice sequences are an apt model for the futurity of the ordinary notion of time, since both are characterized by openness.

But can choice sequences be related to primordial temporality as well? In order to make this connection, we need to link the choice sequence's futurity with the notion of limitedness, that is, with finitude. Here an additional concept must be introduced. There is indeed one context in which choice sequences are characterized by limitation. This characterization is not developed by Brouwer explicitly, but, as Carl Posy has argued, it is required by his position, and something Brouwer himself asserts suggests awareness of this requirement. I will first examine the context that gives rise to the said characterization, and then indicate the parallels between this characterization and Heidegger's account of time.[8]

The context in question is one of Brouwer's proofs for the uniform continuity theorem, which states that every function defined on the real numbers (i.e., every total function) is continuous. This theorem is false in classical mathematics, where the continuum is a set of fully-defined points. One version of the uniform continuity theorem states that "Any discontinuous real-valued function is not total."[9] This can be proved as follows. Assume a discontinuity in the following way: $f(x)=2$ for any $x<½$ and $f(x)=3$ for any $x\geq ½$. Then the value of the choice sequence 0.49999 … with the rule of free choice is not always determined. If at some point we choose a digit other than 9, then the value the function assigns to the number will be 2, because no future choice will change the fact that the number is definitely less than ½. If we choose 9, the value of the function is not necessarily 3, since there is always the possibility that at a subsequent stage a different digit will be chosen, in which case the value of the function will be 2. To prove his theorem, Brouwer requires that the value of the choice sequence be indeterminate, as this will prove that if a function is not continuous, it is not completely defined. But this requirement can be met only if the choice made continues to be 9, though the option of choosing otherwise remains open.

This result is paradoxical, since if 9 is indeed always chosen, then the value of the choice sequence is determined. Posy (2000, 202) describes this situation as "forced indeterminacy."[10] In situations of forced indeterminacy, we must continue to

---

[8] The next paragraph is taken, with minor changes, from Roubach 2008, 65–66.

[9] This proof appears in Brouwer 1927. My reconstruction and the conclusions drawn from it are based on Posy 2000.

[10] Posy uses an analysis of the Hangman paradox to help articulate this notion.

choose a certain digit in order to maintain a series' indeterminacy. According to Posy, the state of forced indeterminacy captures a basic tension in the intuitionistic position. It characterizes Brouwer's position with respect to the notion of infinity (Posy 2000, 206–207), and also lies at the heart of his approach to the continuum. For it is precisely the indeterminacy of the choice sequence that makes it possible for Brouwer to successfully steer clear of two outcomes he seeks to avoid: reduction of the continuum to a set of determinate points, on the one hand, and restricting the arithmetization of the continuum to points that can be defined by rule-governed series and finite choice sequences, on the other.[11]

How is the notion of forced indeterminacy connected to Heidegger's notions of time? The answer requires two steps. I will first argue that primordial temporality, and in particular, Dasein's Being-towards-death, can be described in terms of forced indeterminacy. The second step will be to show the affinities between, on the one hand, the relation between ordinary time and primordial temporality, and, on the other hand, the relation between Brouwer's arithmetized notion of the continuum and forced indeterminacy.

Primordial temporality exemplifies forced indeterminacy since every life has a built-in indeterminacy that is maintained as long as one is alive. Death is always a possibility the realization of which leads not only to the end of a life as a demarcated period of time, but also to the end of any indeterminacy regarding that life. Upon death, the story of our life becomes completely determined. By not dying (for example, by not choosing death), we maintain indeterminacy in our life. In the case of the intuitionistic continuum, maintaining the indeterminacy of the value of the function requires a choice that is, in a certain respect, forced. In other words, maintaining indeterminacy requires limitation of choice.

Given the general affinities, listed above, between Brouwer's intuitionism and Heidegger's position in *Being and Time*, we can interpret the relation between infinite time, viz., time composed of an infinite series of 'now's, and primordial temporality, in terms of the relation between the arithmetization of the continuum and forced indeterminacy. Heidegger's view as to the former relation, we saw, is straightforward: the countedness of ordinary time is rooted in primordial temporality.[12]

Brouwer, we said, rejects the idea that the continuum can be reduced to a set of numerically-characterized points. Only points that can be constructed are acceptable, but these points cannot generate the notion of the mathematical continuum, where every point is determinate. Brouwer's position is not specifically related to time as a continuum, but Michael Dummett (2000) has put forward a Brouwerian approach to the question of the reducibility of time to a set of instants. Dummett begins by rejecting the possibility of discontinuous change. Instead of discontinuous

[11] "Forced indeterminacy," as Mark van Atten pointed out (personal communication, 2012), is not Brouwer's own explicit position. Posy's argument is that in effect, Brouwer's approach leads to this position due to conflation of internal and external perspectives; see Posy 2008. Brouwer's actual formulation is "reserving the right (*die Freiheit vorbehalten*) to determine" (Brouwer 1927, 62).

[12] The remaining paragraphs in this section are based on Roubach 2008, 67–68.

change, he proposes a conception of time on which instants cannot be separated from their surrounding interval, a conception he calls "fuzzy realism." Dummett then offers an intuitionistic model of this conception of time that is based on the notion of choice sequences. By recourse to choice sequences, we are able to narrow the intervals, but we cannot eliminate them completely. Dummett (2000, 515) concludes that "time is only notionally composed of instants, not actually so: instants are unattainable theoretical limits to the process of dissection."

Dummett's approach to the question of the reducibility of continuous time to a set of instants can, I suggest, be applied to the interpretation of Heidegger's stance on the dependence of infinite time on primordial temporality. The procedure for determining moments of time with complete exactitude, that is, the process of generating a series of digits so as to reduce the interval and arrive at determinate points of time, has no limit. The characterization of this process as limitless can be articulated in terms of Posy's notion of forced indeterminacy, but also in terms of Heidegger's notion of primordial temporality.

Let me recapitulate my argument. Given the argument made by Dummett, among others, for rejecting the conception of time as a continuum of instants, and the fact that Brouwer's choice sequences are invoked in describing the process by which instants of time are determined, the notion of forced indeterminacy can be interpreted as helping to explain why the very possibility of formulating the notion of infinite time that is composed of instants is premised on something like Heidegger's notion of finite temporality.

Heidegger's notion of death as the horizon of Dasein's possibilities and the notion of forced indeterminacy are not, of course, identical. Heidegger conceives of death as finite, whereas forced indeterminacy is not necessarily finite. As we saw above, it applies to infinite series as well. But this difference is not crucial, since Heidegger's notion of finitude is not finitude in the sense of finite quantity, but in the sense of limitation, of possibilities that cannot be realized. The profound affinity between the two notions, namely, Heidegger's notion of death and the notion of forced indeterminacy that emerges from Brouwer's approach, is rooted, in my opinion, in the almost-identical roles they play in the respective philosophies. Both Heidegger and Brouwer are seeking to provide a limit that is neither internal nor completely external.

10.3 Conclusion

The proposed argument for Heidegger's thesis of the primacy of primordial temporality over other forms of time offers an opportunity to revisit the question of time consciousness and its relation to time as ordinarily conceived. This problem is usually addressed on the assumption that the two are distinct: ordinary time is taken to be independent of our consciousness of it. Heidegger claims that the two cannot be separated, and are connected via Dasein's primordial temporality. As noted above, and in Roubach (2008, 79), Paul Ricoeur argued that Heidegger did not succeed in establishing the dependence of ordinary time on primordial temporality. He concluded that ascribing autonomy to the ordinary notion of time was unavoidable. This led Ricoeur to introduce his own notion of "narrated time" as a possible bridge between objective–physical time and subjective time. If, as I have argued, Ricoeur's critique of Heidegger should be rejected, then the path is open for rethinking the relationship between conscious time and objective time, a relationship in which primordial temporality plays a key role.

More generally, the continental tradition is often identified with critique of the mathematical representation of time. I have tried to show that this dichotomy should not be sustained, and that philosophical reflection on mathematics can advance our understanding of Heidegger's thought.

References

Becker, O. 1923. Beiträge zur phänomenologischen Begründung der Geometrie und ihrer physikalischen Anwendung. *Jahrbuch für Philosophie und phänomenologische Forschung* 6: 385–560.

Becker, O. 1927. *Mathematische Existenz. Untersuchungen zur Logik und Ontologie mathematischer Phänomene*. Halle: Max Niemeyer.

Blattner, W. 1999. *Heidegger's temporal idealism*. Cambridge: Cambridge University Press.

Brouwer, L.E.J. 1927. Über Definitionsbereiche von Funktionen. *Mathematische Annalen* 97: 60–75.

Dummett, M. 2000. Is time a continuum of instants? *Philosophy* 75: 497–515.

Gallagher, S. 1998. *The inordinance of time*. Evanston: Northwestern University Press.

Heidegger, M. 1962. *Being and Time*. Trans. J. Macquarrie and E. Robinson. Oxford: Blackwell.

Heidegger, M. 1984. *The Metaphysical Foundations of Logic*. Trans. M. Heim. Bloomington: Indiana University Press.

Heidegger, M. 1988. *The Basic Problems of Phenomenology*. Trans. A. Hofstadler. Bloomington: Indiana University Press.

Heidegger, M. 1992. *The Concept of Time*. Trans. W. McNeill. Oxford: Blackwell.

Heidegger, M. 1997. *Plato's Sophist*. Trans. R. Rojcewicz and A. Schuwer. Bloomington: Indiana University Press.

Heidegger, M. 2007. *Basic Concepts of Ancient Philosophy*. Trans. R. Rojcewicz. Bloomington: Indiana University Press.

Posy, C. 2000. Epistemology, ontology, and the continuum. In *The growth of mathematical knowledge*, ed. E. Grosholz and H. Breger, 199–219. Dordrecht: Kluwer.

Posy, C. 2008. Brouwerian infinity. In *One hundred years of intuitionism (1907–2007)*, ed. M. van Atten, P. Boldini, M. Bourdeau, and G. Heinzmann, 21–36. Basel: Birkhäuser.

Ricoeur, P. 1988. *Time and Narrative*, vol. 3. Trans. K. Blamey and D. Pellauer. Chicago: University of Chicago Press.

Roubach, M. 2008. *Being and number in Heidegger's thought*. London: Continuum.

Chapter 11
The Passive Syntheses of Time

Philip Turetzky

Abstract In his later work Husserl explored a genetic phenomenology which underpinned the static phenomenology of his earlier work. Genetic phenomenology explored the genesis of the possibility of judgments as arising out of the fundamental horizon of temporality occurring within primary passivity. This project was taken up by Deleuze in the exploration of a more comprehensive and detailed genesis of individuation and generic difference. In doing so Deleuze described three passive syntheses of time which he presents in a regressive order from conditioned to condition. However, Deleuze's exposition of the most fundamental of especially the third passive synthesis of time is notoriously under described. This essay presents an experimental description which attempts, by reversing the order of exposition of the three syntheses and describing them in the order of genesis, rather than the order of the regress of conditions, to produce a more complete explication of the third passive synthesis and so to account for its genetic dynamism.

Keywords Constitutive paradoxes • Genetic phenomenology • Givenness • Habitus • Immanence • Intensive difference • Passive synthesis • Phenomenological reductions • Porphyry's tree • Transcendental empiricism • Univocity of being

The rubric "continental philosophy" is one of those ragbag terms which, while concealing much and revealing very little, blurs significant differences. The term seems to have arisen among analytic philosophers to whom the various European traditions of thought outside of the analytic tradition, which stemmed from Frege and Russell, seemed to be of a piece.[1] In Europe, however, different philosophical traditions hark back to different ancestors who have bequeathed different problems, methods, and concepts to their philosophical descendents. Different and often

[1] On the confusions inherent in the distinction between analytic and continental philosophy see Bernard Williams "What Might Philosophy Become?" in *Philosophy as a Humanistic Discipline*, 2006, pp. 200–201. I thank Adrian Moore for reminding me of this passage.

P. Turetzky (✉)
Philosophy Department, Colorado State University, Fort Collins, CO 80523, USA
e-mail: turetzky@colostate.edu

Y. Dolev, M. Roubach (eds.), *Cosmological and Psychological Time*, Boston Studies in the Philosophy and History of Science 285,
DOI 10.1007/978-3-319-22590-6_11

sharply contrasting traditions descend from Kant, Hegel, Marx, and Nietzsche, while the influences of Frege and Russell have, until recently, been minimal. This ancestry, for all that, is neither simple nor does it preclude many overlappings and linkages that testify to the complex interrelated flow of the philosophical tradition. Analytic philosophy, for instance, has some of its roots in Kantian and Neo-Kantian genealogical lines, and often seems to swing between Neo-Kantian problems, translated into problems concerning the philosophy of language, and problems (which descend from Locke) in the philosophy of the natural sciences, and which also cross – albeit with often divergent lines of development and differences in emphasis – many of the lines between traditions. So, for example, concerns about the nature and status of the natural sciences arise in different ways in different traditions. To mention only a couple of notable examples, besides the concerns of analytic philosophy of science, critical theory explores the scope and limits of natural scientific rationality, and phenomenology engages both the cognitive and evaluative significance of the natural sciences as in Husserl's final work.[2] Notwithstanding the diversity of these traditions, some traditions in Europe share an ancestry with analytic thought in empiricism, and particularly evoking the influence of Hume.

Hume's challenge to dogmatic thought, as well as forming part of the influence of the British empiricists on analytic philosophy, heavily influenced both Husserl and Deleuze. While Husserl frequently took Hume to task for his psychologism and his consequent failure to escape naturalism, he also praised Hume for being the first to apprehend the transcendental problem: the problem of the conditions under which something can be given, i.e. show itself as it is.[3] What Husserl saw in Hume was an attempt at a completely immanent description of all that shows itself as it shows itself without recourse to anything that could not show itself. Husserl credited Hume with the discovery of the problem of disclosing an immanent constitution and genesis of every object that can show itself insofar as it shows itself, including the constitution of the transcendental field (which Husserl misleadingly calls transcendental subjectivity) itself. Likewise, Deleuze's first book was an exposition of Hume's thought which credited Hume with excising all transcendent objects by rejecting the Cartesian and Lockean notion that the contents of the mind were mere representations of things outside of the mind, things that could not, themselves, be given nor constituted within experience.[4] And like Husserl, Deleuze credits Hume with originating the problem of genesis – of attempting to clarify how, beginning

[2] Husserl, Edmund (1976), *Die Krisis der europäischen Wissenschaften und die tranzendentale Phänomenologie* (*Husserliana* [HUA] VI), mistranslated into English as *The Crisis of European Sciences and Transcendental Phenomenology*, trans. David Carr, Northwestern University Press, Evanston, 1970. The mistranslation in the title introduces an ambiguity that is not in the original, viz. the crisis in the European Sciences is not also a crisis in transcendental phenomenology. There are also many problems with the translation of the text as well.

[3] Husserl, Edmund, *Formal and Transcendental Logic*, trans. Dorion Cairns, Martinus Nijhoff, The Hague, 1969. pp. 256–257 (HUA XVII). On Husserl's assessment of Hume see also HUA VII pp. 152–182. On the relationship between Hume's work and Husserl's see Richard T. Murphy, *Hume and Husserl: Towards Radical Subjectivism*, Martinus Nijhoff Publishers, The Hague, 1980.

[4] Deleuze, Gilles, *Empiricism and Subjectivity: An Essay on Hume's Theory of Nature*, trans. Constantin V. Boundas, Columbia University Press, New York, 1991, pp. 30–31. See also

with what Hume called impressions and operations of the mind, not only things which show themselves but also the subject itself can be constituted immanently. Both Husserl and Deleuze eschew all transcendencies, except insofar as such objects can show themselves immanently. It is in these two ways, abstaining from all appeal to anything that cannot be constituted immanently and focusing on problems of genesis, that their empiricism descends from Hume's and diverges from the representational lines pursued by much of analytic philosophy.

It is, however, difficult – as Hume exemplified – to be an empiricist, i.e. it is difficult not only to adhere solely to that which is given, but to describe that which is given, itself, as it is given without appeal to anything that could not in principle itself be given. Husserl performs what he calls reductions, which effectuate both a suspension of the natural attitude and make possible explications of sense, so as to be assured that descriptions of that which is given include only that which is given insofar as it is given. What Husserl calls the natural attitude encompasses both the attitude of the natural sciences, which posit explanations in terms of transcendencies that cannot show themselves but which are deduced theoretically, and the attitude of everyday life insofar as that attitude assumes that objects can stand on their own, i.e. are substances.[5] Accordingly, the reductions exclude all possible transcendencies and all possible substances.

Husserl often treats the suspension of the natural attitude in two stages in which the phenomenologist refrains, firstly, from accepting objects as existing or not existing as anything other than that which can actually or possibly be intended to in a reflectively experienced mental life. This, however, does not mean that those objects are constituent parts of any mental life, but only that they are to be considered only insofar as they are actually or possibly given through some mode or modes of givenness. This refraining or epoché, if sustained, effects a psychological phenomenological reduction, a reduction of the world to an object (*Gegenstand*[6]) of consciousness. In addition, a second epoché can be performed in which the phe-

---

pp. 86–87 where Deleuze proposes the problem belonging to empiricism as asking, "…how is the subject constituted in the given?"

[5] Here substance merely means that which can stand on its own, or in Descartes phrase, "substance, or a thing that can exist independently" *Meditations on First Philosophy* in *The Philosophical Writings of Descartes*, Vol. 2, translated by John Cottingham, et al., Cambridge University Press, 1984, p. 30. Phenomenologically there are no substances; neither spatial objects, nor mental life, nor even the transcendental field itself stands on its own (in the case of the transcendental field, as we shall see, the field is necessarily open due to its temporality).

[6] There is an important distinction that Husserl marks by the terms *Gegenstand* and *Objekt*. The latter refers to all entities, beings, toward which specifically objectivating acts can be directed (including spatial objects, states of affairs, etc.), while the former more inclusive term refers to anything which can be the subject of true predications. So, *Gegenstanden* include phenomena such as temporal phases, values, and even impossible things like round squares. Husserl's translators have too often blurred this crucial distinction by translating both as "object" (see note 2 above) For more on this distinction see "Husserl's Inaugural Lecture at Freiburg im Breisgau (1917)" in *Husserl: Shorter Works*, edited by Peter McCormick and Frederick Elliston, University of Notre Dame Press, Notre Dame, Indiana, 1981. On Object and objectivating attitude see Robert Welsh Jordan (2008c) http://lamar.colostate.edu/~rwjordan/Notes/W-NotesHUS.HTML#E_objects_[Gegenst%E4nde]_and_objectivating_[Vorstellen]. In this essay, the word "object" will be used to stand for *Gegenstand* unless explicitly noted otherwise.

nomenologist refrains from accepting this reflectively experienced mental life, itself, as a substance. This refraining or epoché, if sustained, effects a transcendental phenomenological reduction. However, the term "reduction" is misleading in this case in that the transcendental reduction widens the field of investigation to include anything of which there can be any consciousness at all, including this reflectively experienced mental life itself (and, notably, including the natural attitude). What is disclosed after performing the reductions is a field in which objects, mental life, and all their characteristics are considered only as senses, i.e. only as something meant and only within the boundaries within which they can be meant.[7]

The point of the transcendental reduction, then, is to secure the phenomena, and the phenomena comprise that which can be given as it is given, or any actual or possible object toward which consciousness may be directed, in other words the transcendental reduction opens a transcendental field comprising the nexus of intentionalities.[8] Phenomena include anything whatsoever of which one can be conscious, i.e. intentive toward, in any manner whatsoever, including intentionalities themselves, toward which one can be intentive in reflection. Intentionality is not a relation between two things which could stand on their own (substances), but is an inseparable unity from which mental life and objects can only be separated by abstraction from phenomena. The world and consciousness belong together inextricably. Consequently, the traditional notions of subject and object apply only to abstractions which must be constituted and generated, within the transcendental field.

Intentionalities are polar unities; the two poles are that which is given as it is given, the noema, and the manner of givenness, the noesis. So, for example, in a perception of a spatial object the spatial object insofar as it is given is the noema and the act of perceiving is the noesis. Moreover, some aspects of the object are actually given, i.e. given in intuition,[9] while other aspects are given only as possibilities or

---

[7] Husserl expresses this in"…the *principle of all principles: that every originary presentive intuition is a legitimizing source of cognition*, that *everything originarily* (so to speak, in its "personal" actuality) *offered* to us *in "intuition" is to be accepted simply as what it is presented as being*, but also *only within the limits in which it is presented there.*" Edmund Husserl, *Ideas Pertaining to a Pure Phenomenology and to a Phenomenological Philosophy*, first book, translated by F. Kersten, Martinus Nijhoff Publishers, The Hague, 1983, p. 44 (HUA III).

[8] For a clear and detailed explication of intentionality see Robert Welsh Jordan's (2008b) *Notes on the Intentionality of Consciousness* on his website at http://lamar.colostate.edu/~rwjordan/Notes/NotesInt.HTML.

[9] See footnote 7 above. The horizonal structure of experience as including the openness of objects onto their possibilities is a theme in all of Husserl's writings beginning with Edmund Husserl, *Logical Investigations*, trans. J. N. Findlay, Vol. 1 Routledge, New York, 2001a, pp. 176, 199–200 (HUA XVIII). For an example from his later work see, Edmund Husserl, *Erfahrung und Urteil.* Untersuchungen zur Genealogie der Logik. Redigiert und heruasgegeben von Ludwig Landgrebe. London, Eng.: Allen and Unwin, 1939, Einleitung § 8. Husserl clarifies the apperception of something not itself given, including possibilities, as follows: "Apperceptions are intentional lived-experiences that are conscious of something as perceived which is not self-given in these lived-experiences (not completely); and they are called apperceptions to the extent that they have this trait, even if in this case they also consciously intend what in truth is self-given in them."

potentialities of that object, but not themselves intuitively given. Instead these aspects are apperceived as belonging to that object. So, in perceiving a spatial object, aspects of the object such as its far side are not intuitively given, but are apperceived as belonging to the noema. These apperceptions are forms of empty intendings in which something is meant but not itself given. This means that objects are given along with their specific horizon of meaning or noematic sense; we live in and are given a meaningful world, a world in which we, for example, perceive spatial objects, other persons, cultural objects such as artworks, etc. not only as they are given intuitively, but also insofar as their possibilities and potentialities are given as well. Since the transcendental reduction suspends the attitude of the natural sciences, it becomes possible to describe experience as meaningful without recourse to transcendent causes nor to representational theories intervening and introducing transcendencies which cannot themselves be given.

Describing that which is given as it is given requires, then, that the sense of that which is given be explicated – that the sense of that which is given is itself given already means that that sense can be explicated. The sense, or what is meant, i.e. intended, can be the sense of this particular object or can be the sense of either formal or material universals,[10] i.e. *eide*, exemplified by any given object. Moreover, *eide* can themselves be given as objects which are neither themselves constituent parts of the noemata which exemplify them, nor are they immanent to any consciousness which may intend them. Even though *eide*, as such are not constituent parts of anything actual, they can be given, and given adequately; this is only possible because they are founded upon particular objects which are actually given. In each of the stages of reduction, it is possible to perform what Husserl calls an *eidetic* reduction, i.e. to bring to givenness, *eide* exemplified by each noema and which explicate the sense of that noema in its actual or possible givenness. This understanding of *eidetic* objects contrasts phenomenology sharply both with seventeenth and eighteenth century empiricist, and all Kantian and Neo-Kantian approaches to universals, which treat universals as products of active subjectivity (what Kant calls spontaneity) and with those versions of analytic philosophy which treat universals as somehow linguistic.

Deleuze is often mistakenly thought to be a naturalist[11] because he, like Heidegger, does not perform the transcendental reduction explicitly, but insofar as what he calls transcendental empiricism seeks to remain entirely within the immanence of the transcendental field, his work, in effect, remains within the transcendental reduction. Deleuze writes, "Although it is always possible to invoke a

Edmund Husserl, *Analyses Concerning Passive and Active Synthesis: Lectures on Transcendental Logic*, trans. Anthony J. Steinbock, Kluwer Academic Publishers, 2001b, p. 624 see also 624 ff. (HUA XI).

[10] op. cit Robert Welsh Jordan (2008d) http://lamar.colostate.edu/~rwjordan/Notes/W-NotesHUS.HTML#E_formal_essences,_superordinate_to_material_essences.

[11] See for example, Manuel Delanda, *Intensive Science and Virtual Philosophy*, Continuum International Publishing Group, 2002. For a treatment of Deleuze as phenomenologist specifically focusing on the passive syntheses of time see Joe Hughes, *Deleuze and the Genesis of Representation*, Continuum, 2008.

transcendent that falls outside the plane of immanence, or that attributes immanence to itself, all transcendence is constituted solely in the flow of immanent consciousness that belongs to this plane. Transcendence is always a product of immanence."[12] Moreover, Deleuze's use of concepts derived from the natural sciences, as well as those concepts derived from the arts and mathematics are part of his philosophical project of creating concepts and not an appeal to natural science as a justification for or as a naturalistic constraint on philosophical conclusions; he does not adopt these concepts for any but philosophical purposes.[13]

In addition, Deleuze's position is often thought to be at odds with Husserl's because Deleuze is critical of any position centered around the subject and subjectivity, and in addition because Deleuze argues against characterizing transcendental conditions as conditions of possible rather than of real experience. The former objection misunderstands the character of the transcendental field in both positions. For Husserl, despite his talk of transcendental subjectivity and of a transcendental ego, both the object and the subject must be constituted within a transcendental field which is prior to the unity of either objects or subjects and which belongs neither to the subject nor the object, but to a pure immanence within which subject and object must be constituted.[14] We shall see this in more detail below in discussing passivity and passive syntheses. Husserl's tendency to write in terms of subjectivity and to designate the transcendental field as a transcendental ego partly results from his critique of naturalism, and partly results from the philosophical project that guides his work. The critique of naturalism entails, for Husserl, the transcendental move from natural objects to considering such objects only under the conditions through which such objects can be given. The transcendental reduction arrives at these conditions by establishing the nexus of intentionalities, with their irreducible noetic-noematic structure, as giving objects only insofar as they are given through various modes of givenness. Moreover, Husserl's project, to describe the structures and unity of the transcendental field and how objects are constituted within that unity, tends (partly for historical reasons and partly as a result of the order in which Husserl's thought developed) to emphasize – in contrast with naturalism – *how* objects are given, and along with that the unity of the subjective poles of the nexus of intentionalities.

---

[12] Gilles Deleuze, "Immanence: A Life" in *Pure Immanence*, Zone Books, New York, 2001, p. 31, see also the footnote, p. 33 where Deleuze cites Husserl's agreement with this claim.

[13] For Deleuze's detailed discussion of the differences between the projects of the natural science, which create functions, and philosophy, which creates concepts, see Deleuze, Gilles and Guattari, Felix, *What is Philosophy?*, Columbia University Press, 1994. In a remark quoted in the translator's preface to Deleuze, Gilles and Guattari, Felix, *A Thousand Plateaus: Capitalism and Schizophrenia*, trans. Brian Massumi, University of Minnesota Press, Minneapolis, 1987 (a work pervaded by the use of concepts derived from the natural sciences, the arts, and mathematics), and which can be applied to all of his work, Deleuze said that he was doing "philosophy and nothing but philosophy." See Gilles Deleuze, interview with Catherine Clement, *L'Arc*, no. 49 (revised ed., 1980), p. 99.

[14] Edmund Husserl, *Introduction to Logic and Theory of Knowledge: Lectures 1906/07*, trans. Claire Ortiz Hill, Springer, Dordrecht, The Netherlands, 2008, p. 243.

With respect to Deleuze's concern with the conditions of real experience, that concern focuses on issues of genesis which were only partly taken up in Husserl's later work, especially issues regarding the genesis of individual objects. It was late in Husserl's development that he came to focus on and to fully understand the importance of genetic issues on which Deleuze's philosophical project focuses, and even then Husserl only developed the genetic problematic in a narrow and limited way. By the 1920s, although he had been aware of the need for a genetic analysis, even as early as the 1905 time consciousness lectures, Husserl came to realize that the static analysis that had been in the forefront of his concern since the *Logical Investigations* (an analysis that sought to describe the noetic acts and noematic objects, the relations between empty and intuitively fulfilled intendings, and the *eidetic* structures and possibilities exemplified by objects and acts) was neither basic nor adequate, and that a genetic analysis underlay these static analyses. Genetic analyses, Husserl came to think, were more concrete while static analyses remained too abstract.[15] A genetic analysis requires that the unity of perceived objects (which founds the givenness of *eidetic* structures, so that Husserl's genetic analyses bear on judgments concerning individual objects which is all we need to be concerned with here) and the unity of the transcendental field itself requires a genetic history, a temporal becoming and development of the constitution of objects within the transcendental field and of the transcendental field itself.[16] The transcendental field, then, is essentially temporal, a field of becoming both of the objects which are themselves unities of becoming and of the field itself as a unity of becoming, i.e. a unity of ceaseless genesis.[17] Genetic phenomenology, then, is the phenomenology of the intrinsic becoming of objects in time, and the becoming of time itself since the transcendental field is ineluctably temporal.

Genetic phenomenology, for Husserl, seeks to describe three layers of experience: genesis within active consciousness, i.e. consciousness attentive towards objects and in which the ego performs rational acts (active consciousness is often but should not be confused with consciousness as such[18]), genesis within passive experience, which includes affection, apperception, kinaesthesis, etc., and in between, genesis of active forms of consciousness from passive forms of consciousness such that all active experience originates out of passive experience and passive processes.[19] In this essay, we shall focus, primarily, on the sphere of passivity and on the generative dynamics of passive processes. Unsurprisingly, genetic analysis and emphasis on passivity came to the fore together in Husserl's work, since the two are interrelated; passive genesis comprises the most primordial immanent pro-

[15] On the development of Husserl's understanding of the relationship between genetic and static phenomenology see Husserl 2001, pp. 624–645 and Steinbock's introduction to this volume, pp. xxviii–xxxviii.

[16] Ibid., p. 644 "…attending to constitution is not attending to genesis, which is precisely the genesis of constitution…"

[17] Ibid., p. 635.

[18] Ibid., p. 92 ff.

[19] Ibid., p. 631.

cesses – processes of the becoming of sense, as well as the middle level of the genesis of active from passive experience. Passivity also comprises aesthesis as founding judgments and thereby all cognitive experience; passivity supplies the conditions for the possibility of subjectivity.[20] The most fundamental passive processes are the passive temporal syntheses, since all that is immanent is encompassed by temporality. Temporal modes of givenness generate succession, coexistence, forms of connection, individuation of objects, and the constitution of the identity and unity of all possible objects. On the noematic side, individuation of objects, unities of identity and heterogeneities of opposition are generated through temporal syntheses. Passive consciousness is that sphere in which the ego does not actively participate in the constitution of sense. Passive processes, however, retain intentional structure; so even sensations must be constituted, insofar as they are unities, by syntheses of intentions.[21] There are strivings, drives, and tendencies towards objectlike (*Gegenständlichkeit*) formations, in which the ego does not actively participate. Passive experience is pre-reflective, pre-given (with respect to active consciousness), pre-individual, and pre-predicative, and so does not present objects as complex individuated correlates of objectivating acts, since objectivating acts require activity on the part of the ego. That which is pregiven in passivity is experienced affectively through its prominences which affect and exert attractions within the transcendental field without being grasped attentively as exerting this allure; in this sense the transcendental field as a field of genesis is a field of forces, albeit not transcendent causal forces which are ruled out by the reductions. The passive sphere of experience is as close as phenomenology can come to a conception of the unconscious, and both Husserl and Deleuze refer to it as such.[22] This unconscious includes a region of instincts and habits (some of which may have once been actively acquired and have become sedimented so as to operate pre-reflectively, in what Husserl calls secondary passivity).

Deleuze, however, pursued the realm of passivity beyond Husserl's inquiries by more completely appreciating the scope of that which is founded on passive processes, and thereby disclosing the modes of genesis, themselves, in a more fundamental way, resulting in an extension and critique of Husserl's account of genesis. Firstly, while Husserl, in accord with his philosophical project, focuses mainly on problems which arise concerning activity, the genesis of activity out of passivity, and in particular, the genesis of judgment out of aesthesis, Deleuze extends the problematic to include the genesis of the whole context of Porphyry's tree. While Husserl focuses solely on the genesis of the middle more traditional portion of the tree in which genus-species relations (identity in the concept and opposition of predicates) are presupposed to judgments, Deleuze is concerned with the genesis of

---

[20] Ibid., p. 163.

[21] It is partly due to Husserl's realization that the constitution of sensations must be intentional that he came to focus on passive genesis. See ibid. translator's introduction, pp. lv–lvi.

[22] See Husserl, op. cit. p. 201 and passim. and Gilles Deleuze, *Difference and Repetition*, Columbia University Press, New York, 1994, pp. 106 ff. and p. 174.

the tree and its context or what he calls representation[23] which, in addition to genus-species relations, includes both the very possibility of categories and generic differences as such and also individual differences represented by way of resemblance in perception. These lacunae are crucial since Husserl's account of passive processes presupposes resemblances as simply given and treats differences only in terms of oppositions between predicates which fall under genera in which they share an identity within those genera. In this way, Husserl takes both individual and generic difference for granted which amounts to accepting unexamined presuppositions. Yet, Husserl's own prescription forbids phenomenology from unquestioningly accepting any such presuppositions.

	Being			
	Substance, etc.* (the categories)		generic difference (highest determinable concepts)	ANALOGY of judgement
Corporeal		Incorporeal	specific difference	
	Body		subaltern genus	
Animate		Inanimate	specific difference	IDENTITY in the concept
	Living thing		subaltern genus	
Sensitive		Insensitive	specific difference	
	Animal		subaltern genus	OPPOSITION of predicates
Rational		Nonrational	specific difference	
	Human		infima species (smallest determined concepts)	
Socrates	Plato	etc.	individual difference (individuals)	RESEMBLANCE in perception

*Aristotle's ten categories: substance, quality, quantity, relation, place, time, position, state, activity, passivity.

Porphyry's Tree: enhanced version[24]

Deleuze, in effect following Husserl's prescription, questions these presuppositions and argues that the whole of representation must be generated out of univocal being and that being as univocal must be understood as difference in itself.[25] Being is neither reducible to substance nor merely, as Husserl seemed to think, applicable to what is given in objectivating acts. In this, Deleuze follows Heidegger in raising the question of being with respect to the transcendental field itself, a point which

[23] Deleuze, op. cit. 1994, pp. 34–35, and passim.

[24] from Daniel W. Smith "The Doctrine of Univocity: Deleuze's Ontology of Immanence" in *Deleuze and Religion*, ed. Mary Bryden, Routledge, London, 2001, pp. 167–183.

[25] Deleuze, ibid., pp. 28–69. On univocality see also Smith op. cit. Deleuze couches his critique of these presuppositions by arguing that Husserl presupposes what Deleuze calls common sense and good sense, see Deleuze, ibid., pp. 129–134 and passim.

Husserl did not acknowledge and over which Heidegger broke with Husserl.[26] Being, as Aristotle argued,[27] must be distinguished from a genus, but contrary to Aristotle and working within the transcendental reduction, being must be distinguished from substance as well. Difference in itself is the univocal condition for the whole of representation including analogy within judgments, identity in genera, opposition between predicates, and resemblance in perception. In this sense, difference must be understood not as opposition (of predicates) within identities provided by higher genera nor even as individual differences due to contrasts and resemblances, but as intrinsic and intensive difference. As intrinsic to all phenomena (including all actual, potential, and possible phenomena), difference is univocal, having only one sense, especially since it is the genetic condition for the categories. As intensive, difference furnishes, as we shall see, the generative power of the temporal syntheses.

Instead of treating sense as having the form of a predicate (in the form neither of genus nor species), and instead of unquestioningly accepting the identity of objects in general or presupposing the order of a sedentary hierarchy from most differentiated to the least, sense and the genesis of sense must be understood as having the form of an event, as expressed by the verb rather than the predicate, and hence as ineluctably temporal.[28] Deleuze writes that "the event is sense itself."[29] Three aspects delineate the event, a totality, an empty form, and a series; these three compose time as genetic and structure what Deleuze designates as the third passive synthesis of time. Deleuze presents three passive syntheses that generate and constitute time and Deleuze (and most others who write on the topic) usually presents these syntheses ordinally and regressively from conditioned to condition, so that the second synthesis is the condition of the first and the third is the condition of the first and second. In the order of genesis, however, the third synthesis, the Event, comes first, so we will begin with it.

Before beginning that exposition, note that Deleuze only discusses syntheses in relation to the passive syntheses of time in *Difference and Repetition*. In *Anti-Oedipus*, Deleuze and Guattari do indeed discuss syntheses in the order of genesis, but not explicitly in terms of the genesis of time, and there the order of the syntheses is different[30]; *Anti-Oedipus*, for the most part follows the lines of dis-

---

[26] See Steven Galt Crowell, *Husserl, Heidegger, and the Space of Meaning: Paths Toward Transcendental Philosophy*, Northwestern University Press, Evanston Illinois, 2001, pp. 167–181.

[27] Aristotle *Metaphysics*: Book III (B) 998b 4 ff.

[28] Gilles Deleuze, *The Logic of Sense*, The Athlone Press, London, 1990, pp. 96–99. On the transformation of the event into predicative form see p. 111 ff.

[29] ibid., p. 22.

[30] It is worth noting, however, that much the same structures and dynamics are described in *Anti-Oedipus* as found in *The Logic of Sense*. In particular, what in *Anti-Oedipus* are called desiring-machines, which produce flows on one side and break flows on the other synthesizing the production of production, align with intensity in *Difference and Repetition* and the paradoxical element in *The Logic of Sense* which produces indefinite series on one side and vicious circles on the other.

cussion in *The Logic of Sense*. Unfortunately, even in *Difference and Repetition* the third synthesis of time is notoriously under described and so by necessity much of what follows concerning the third synthesis of time will be experimental, drawing not only on Deleuze's explicit discussion of the passive syntheses of time in *Difference and Repetition* but supplementing that discussion with material from *The Logic of Sense* in which the issue is the genesis of sense. The latter discussion is apposite both because Deleuze treats sense itself as the event, and because in pursuing the positive alignment between Deleuze and Husserl regarding the genesis of time it is still the case that despite his critique of Husserl's treatment of sense in terms of the predicate, Deleuze explicitly regards sense as noematic sense.[31]

As a totality, the Event expresses time, as pure occurrence, i.e. a becoming gathering together all events in a great Event equal to the open whole of time. This open totality of time serves as the limit and condition of immanence.[32] Deleuze connects the great Event of time itself with Nietzsche's conception of the eternal return. For Nietzsche, the world is a chance configuration of forces, not of things, and the eternal return gathers these relations and encounters between forces together into a great Event.[33] The principle of eternal return affirms all events, past and future, as connected within that great Event. Nietzsche, in the persona of his Zarathustra, writes "…are not all things firmly knotted together in such a way that this moment draws after it all things to come? Therefore – itself as well?"[34] Hence, the Event is both adequate to time as an open whole, but is also divided into two unequal parts that draw together the totality of time[35]: that which will be given, i.e. is given only in the form of potentialities, and which is unequal to that which has been actual.

Phenomenology insists on the openness of temporality as a totality and hence of the totality of that which is immanent. Husserl describes this open temporal horizon as an obscure horizon of protentions (intendings of that which is to come) and retentions (intendings of that which has itself been) through which determinate content is inserted into the temporal flow.[36] That which is determinate cannot be separated from this obscure horizon. Moreover, and appositely, all protentions, regardless of their clarity or obscurity are empty in that they can only be directed toward potentialities and possibilities, i.e. directed to those potentialities and possibilities

[31] Deleuze 1990, pp. 20–21.

[32] The Event as a totality serves as a boundary condition on phenomena in accord with the thesis of my book *Time*, 1998.

[33] On the eternal return see Friedrich Nietzsche, *The Will To Power*, Vintage, 1968, 617, Friedrich Nietzsche (2006), *Thus Spoke Zarathustra*, translated Adrian Del Caro, p. 126. pp. 264, 267–272, and pp. 327–333. *The Gay Science*, trans. Walter Kaufmann, Vintage Books, 1974 §§109, and 341.

[34] Nietzsche 1974 p. 270.

[35] Deleuze 1994, p. 89. See also Deleuze 1990, pp. 53–4.

[36] Edmund Husserl, *On the Phenomenology of the Consciousness of Internal Time (1893–1917)*, trans. John Barnett Brough, Kluwer Academic Press, Dordrecht, 1991, p. 89, (HUA X).

themselves even though it is contingent whether any of them become actualized.[37,38] Hence, even though temporality forms a totality, that totality is open onto those potentialities and possibilities and changes continuously as new phenomena become actualized. That something new is continuously produced is intrinsic to the Event making the third synthesis of time the primary synthesis of the future, even though that which is to come is not yet actual and protended possibilities may never be fulfilled.

Something new continuously occurs as the actualization of a potentiality that shows itself when an empty possibility comes to be fulfilled. This means that there is a break or split in time, what Deleuze calls, in a significant allusion to rhythmic structures, a caesura, between the potentialities and empty possibilities of the future and that which has become actualized as moments recede into the past. Time unfolds from this caesura that draws together and differentiates the totality of time formally into its two unequal parts forming the second aspect of the third synthesis of time, the empty form of time. This formal structure of time can only be abstract[39] and individual. It can only be abstract since concrete temporal structure cannot be indifferent to its content whether potential or actual; for there to be temporality at all requires that something be given. This is the sense in which Deleuze, quoting Hamlet, says that, in the third synthesis, time is out of joint, i.e. the empty form of time is abstracted from its content.[40] Furthermore, the empty form of time is not an eidetic object, but is an individual form[41] since the temporal structure of any constituent given in the temporal flow can occur only when that constituent is actualized; and it must occur with its temporal structure whenever it will be or is or was occurring. Its temporal structure, since it is abstracted from this content, has its individual peculiarities which individuate it, and that constituent belongs to the temporal flow even when the constituent does not change.

The caesura is the instant of metamorphosis; the singular transition point between potentiality and actualization. The caesura is not itself a phase, but is a singularity that marks the break between the future and the past. Passing through the caesura that which is potential in the future and has become actual in the past must in the caesura be both at once.[42] The new arises in the caesura which then, like the break

[37] It is contingent that the transcendental field, the field of immanence, exists at all, hence the significance Heidegger attributes to death.

[38] Ibid., p. 309 note. See also Husserl 2001, pp. 129, 170, 365, and 614.

[39] Edmund Husserl, *Cartesian Meditations: An Introduction to Phenomenology*, trans. Dorion Cairns, Martinus Nijhoff Publishers, The Hague, 1982, p. 75, (HUA I), "We can call it furthermore a *formal regularity pertaining to a universal genesis*, which is such that past, present, and future, become unitarily constituted over and over again, in a certain noetic-noematic formal structure of flowing modes of givenness."

[40] Deleuze 1994, p. 88.

[41] Robert Welsh Jordan (2008a) http://lamar.colostate.edu/~rwjordan/Notes/W-NotesHUS.HTML#E_time,_as_the_Universal_form. "The transcendental ego and all of its really inherent constituents (all of which are abstract parts) are individual objects. None of them is something universal [Allegmeines] and none of them is something eidetic. Nothing individual can have constituents, really inherent parts, that would be other than individual."

[42] We shall return to this paradox below.

between rhythmic feet, from which the term originates, draws together potential and actualized in their difference, the caesura acting as a condition of real experience, a condition of the new. The caesura both distinguishes protentional phases of potentialities from retentional phases of real experience, and marks the singularity in which new experiences (inseparably including both objects as given and their modes of givenness) arise and which reverberate over the sense of that which has already been retained and thereby modify potentialities to come. These potentialities can only be protended on the basis of that which has been[43] and are modified with the retention of new phases. The potentialities to come, which Husserl calls problematic possibilities,[44] consist in prominences, remarkable points, which are problematic and indeterminate with respect to whether they will come to be actualized. In this sense, the third synthesis of time generates a rhythmic distribution of singular and ordinary points, accented and unaccented moments. Rhythms, here, are ways purely temporal intervals become grouped together by distributing accented and unaccented moments, ways in which unaccented moments are grouped in relation to an accented one independent of any periodicity.[45]

Each of the three passive syntheses of time contains a past, present, and future series belonging to that synthesis. With respect to the third synthesis, the caesura, as the instant of metamorphosis, serves not only as the basis of the empty form of time, but also and more concretely, differentiates time into a series by distributing events unequally on either side of the caesura. The caesura, Deleuze writes, is itself the present of metamorphosis,[46] the singular point in which something new can arise. Deleuze often describes the temporal series in terms of the self challenged to perform a deed; the past is determined by not being capable of the deed, the present is determined by becoming capable of the deed, and the future is determined by the self being exceeded by the deed.[47] However, the series must be understood on the object side and also more generally in terms of individuation, identity, and the event – of something new arising. Individuations and identities, including those of objects and the self must be generated by the temporal syntheses. The determination of the series by its relation to the instant of metamorphosis allows the series not only to be determined but also to be generated without presupposing the series is already in place.

The past is situated as before the metamorphosis, as containing identities that are not yet capable of transformation, as Deleuze puts the point, not yet equal to the event. For both Husserl and Deleuze, this past belongs primarily to the second synthesis in the form of the past in itself. The present is situated at the caesura, at the point at which something new arises marking the event in which actual doubles potential, individuating objects and transforming identities, in which identities

[43] See the discussion of the second passive synthesis of time, below.

[44] See for only one example, Husserl 2001, p. 83.

[45] See Cooper, Grosvenor and Meyer Leonard B., *The Rhythmic Structure of Music*, The University of Chicago Press, Chicago, 1960, Chap. 1.

[46] Deleuze 1994 p. 89.

[47] Ibid.

become capable of, equal to, the event of their own transformation. So, in Deleuze's example, the outcome of Oedipus' inquiry transforms him so that Oedipus the investigator becomes equal to Oedipus the actor who has done the deed which inseparably transforms him and the world given to him. For Husserl as well, the break between the past and that which has not yet come to be actualized is not an abstraction. In Husserl's terms, the present impressional moment is the distinguishable temporal phase in which something affects the self in passivity: it is only in the present that an affective force can bear upon the self and so for Husserl as well as for Deleuze, affection is "the most universal genetic phenomenon."[48] Affection is a becoming prominent of something in passivity so that it becomes capable of arousing a practical interest on the part of the self, i.e. affection is the individual form of all that can come to have been prominent. Furthermore, the present impressional moment of original affection does not give an object (of an objectivating act), i.e. something with an already established identity, but rather serves as the genetic moment which can result in the constitution of identities and the individuation of objects through other passive syntheses, in particular through the passive associative syntheses which take place in the first passive synthesis of time.

While potentialities and possibilities may only be projected upon that which is to come on the basis of the past in itself found in the second synthesis, this projection on that which is to come differs sharply from the generative future in the third passive synthesis of time, where the future has priority as the fundamental dimension out of which something new can arise. While this is true for both Deleuze and Husserl, Deleuze is more radical in this respect. Husserl tends to emphasize the first two syntheses of time and the projection of past forms onto future possibilities, possibilities which have already been established, while Deleuze, consonant with his wider project pursuing passive processes and disclosing the modes of genesis of all forms of representation (in Deleuze's sense), emphasizes the transformative capacities of the third synthesis. The conditions of the generation of the new include the future as potentiality uncontaminated by anything actual. This means in addition to emphasizing the individuation of objects and the generation of identities, objects and identities not already constituted, the event operates by excess, i.e. by exceeding identities, and because of the potentiality which defines the future, the event transforms objects in their individuation and excludes the coherence of identities in the singularity within which potentialities come to be actualized. Deleuze puts this dramatically, writing:

> As for the third time in which the future appears, this signifies that the event and the act possess a secret coherence which excludes that of the self; that they turn back against the self which has become their equal and smash it to pieces, as though the bearer of the new world were carried away and dispersed by the shock of the multiplicity to which it gives birth: what the self has become equal to is the unequal in itself.[49]

[48] Husserl 2001, p. 184, for Husserl's account of affection see Husserl 2001, p. 216. For an excellent account of these and the following points with respect to the present as the moment of original affection see Andrea Staiti, "The Primacy of the Present: Metaphysical Ballast or Phenomenological Finding?," *Research in Phenomenology* 40 (2010) 34–54.

[49] Deleuze 1994 pp. 89–90.

In this way, the Event returns and that which returns in the Event is difference in itself,[50] differentiation constituting the continuous genesis of the new. The third synthesis of time is the form of change, the element of differentiation, which does not itself change.

But Deleuze does not leave these matters in this undeveloped form; he extends the exposition of the genesis of the new to disclose its most basic genetic structures. This means he, like Husserl, does not countenance the notion that at the root of genesis is a fundamentally unstructured formlessness.[51] Instead, Deleuze explicitly characterizes difference in itself and the structures of genesis by means of intensity and paradoxes.

Time, as it is experienced and as it finds its genesis in the third synthesis, is intensive. When Deleuze says that in the third synthesis the self (identities) become equal to the unequal in itself, he is referring to intensity as difference in itself – difference in itself is intensive difference. Intensities are marked by three characteristics[52]: an internal difference, ordinal distances between differences forming a series, and the envelopment of differences of lower rank in the series by differences of higher rank. The split in time, the sharp distinction between potentialities and the actualities which actualize some of those potentialities, forms a difference, an inequality, internal to time, i.e. internal to and arising out of the caesura. The genesis or actualization of potentiality forms an intensive line extending this inequality in a series of ordinal distances without resolving the inequality.[53] The intensive genesis of time occurs in the caesura and so the series of distances does not take place over an extensive temporal interval,[54] but is orthogonal to the future, present, past series and can serve as the source of its genesis. Since this intensive series does not extend over a temporal interval, it exhibits an envelopment of each difference in the series by that which displaces it. Time, as it is experienced, is a continuous envelopment of the past as new experiences arise.

Enveloped differences are implicated in and implicate the enveloping differences. This ordinal implicative series is a determinate form of being,[55] something

---

[50] Gilles Deleuze, *Nietzsche and Philosophy*, Columbia University Press, New York, 1983, pp. 48 and 189–190.

[51] Even in his discussions of sense arising out of nonsense, Deleuze does not treat nonsense as an undifferentiated irrationality (as something that cannot be explicated), and he explicitly distinguishes the transcendental field, insofar as it consists of differences, from an undifferentiated nothingness, see Deleuze 1994, p. 28. Hence, despite his debt to Bergson, Deleuze, like Husserl, and contrary to what many expositors of Deleuze write, is critical of Bergson's and other forms of vitalism which posit something undifferentiated at the core of time.

[52] While Deleuze discusses intensity throughout his work, the most detailed discussion occurs in Deleuze 1994, pp. 232 ff.

[53] Deleuze (1994, pp. 330–331 note 13) cites Russell (The Principles of Mathematics, Cambridge: Cambridge University Press, 1903, Chap. 31) regarding ordinal distance as an asymmetrical relation that is indivisible by any common factor.

[54] Here we should note that Husserl too affirms that transcendental syntheses do not themselves occur over temporal intervals. See Husserl 1991, p. 381 ff.

[55] Deleuze 1994, p. 237.

positive founding ordinal distances in which the intensities are positive and generative, the series of differences is determined neither by opposition of predicates nor the lack or negation of something already constituted (consequently, Deleuze marks internal difference with the sign of the differential, dx, rather than by not-x). Neither of these can precede the condition of their genesis, since both are aspects of representation in Deleuze's sense. As positive and genetic, difference serves as a transcendental principle, but one that can itself be given immanently, engendering what Deleuze calls a transcendental empiricism in which the conditions of the given can themselves be given, once more aligning Deleuze with Husserl who only countenances conditions that can be given.

Distance is an asymmetric ordinal relation which cannot be divided except insofar as each distance in the series differs from the others, every point in the series being an internal difference. Each degree of the intensive series is heterogeneous and the series ordinal so that it contains no common measure which would determine an equivalence between the distances; each degree is unequal both in itself and in its relation to other degrees. Intensity, then, is prior to both quality and (extensive) quantity; it is "neither divisible like extensive quantity nor indivisible like quality;"[56] it cannot be divided according to some equivalence relation, yet it is not undifferentiated but consists of differences begetting differences. For example, degrees of heat contain an intensive series in which each degree differs from the others such that there is no equivalence relation internal to the series; these differences can only be given a common measure in a dimension external to the series as when degrees of heat are measured extensively for instance in terms of the length of a column of mercury where the extensive length is not internal to the series of heats. Since an intensive series contains no common measure, no equivalence relation internal to it, it is unequal and unequal in itself, as Deleuze puts it, the series cannot be divided "without changing nature,"[57] i.e. the parts of the whole arise from the division and are not of the same sort as the whole nor as the other parts. This means that intensity forms both a series of differences in themselves and a whole, a totality – the totality of the Event, a series that cannot be prolonged without differing from itself. In essence intensity is difference in itself.

While the third synthesis occurs in the caesura, in the moment, i.e. in an instantaneous phase, and does not occur over a temporal interval, it is nevertheless dynamic. The dynamism internal to intensity arises through constitutive paradoxes. These are paradoxes not in the epistemological and negative sense of contrary to opinion, but as the ontological and positive immanent origin of time. Such paradoxes are not errors in nor barriers to thought, but are instead dynamic principles of genesis. Paradoxes, in this sense, are essentially structured so as to generate vicious circles on one side and infinite series on the other. Paradoxes such as Russell's paradox and McTaggart's paradox have this form; vicious circles arise insofar as an element turns back on itself (e.g., a name names itself), while indefinitely expanding series arise when that element becomes displaced (since it is circular for a name to name itself the naming gets displaced to another name that names the first and so

---

[56] Ibid.

[57] Ibid.

on – the circularity of the name that names itself is thereby displaced in a continuous production of sense, e.g. Russell's hierarchy of types).

The dynamism of the paradox arises from a paradoxical element, an element which arises when a constituent turns on itself, doubles itself, and since it thereby becomes continuously displaced in relation to itself traverses an ever expanding series without being able to rest at any point. The paradoxical element has no stable place in the series. A series of empty places becomes continuously traversed and exceeded by an occupant without a place. "It also fails to observe its own identity, resemblance, equilibrium, and origin."[58] The paradoxical element cannot settle into an equilibrium, since it is continuously displaced, and has no identity in that it is always unequal to itself. Since only differences without a common measure belong to the intensive series it structures, it fails to even resemble itself. Hence, it is capable of generating the aspects of representation without presupposing any of them; it embodies a transcendental principle which is impersonal, lacking the form of a subjective identity, and is pre-individual in that it cannot be individuated since it traverses an indefinitely expanding series without occupying a place in the series. And, since it is unable to resemble itself, it is neither general nor universal. Finally, due to its very dynamism, its continuous displacement, the paradoxical element exceeds its origin while manifesting the genetic power to constitute the very series it traverses. The paradoxical element acts as a quasi-cause immanent to its effects which in turn engender further effects.[59]

A series of paradoxes constitute the immanent genesis of time. The first has two aspects which belong to the paradox as such. The initial aspect arises from the structure of the caesura; the moment of transformation includes itself, turns on itself and doubles itself. Nietzsche alludes to this in the passage quoted earlier in which the moment of transformation elicits all that is to come – including this very moment. This moment includes itself insofar as its actuality doubles its potentiality without resembling it; in the caesura the moment is actual on one side and potential on the other. This turning back on itself is the paradoxical element insofar as it takes the form of a vicious circle. The moment repeats itself, but repeats itself only insofar as it differs from itself.[60] The Event includes all events and forms an open totality, the whole of time in an instant; the dimensions of past and future unfold from the caesura. Yet because it differs from itself, is intrinsically unequal, this totality cannot remain closed, but must be displaced in an intensive line that extends each inequality without resolving that inequality, but reproducing it with more intensity generating an indefinitely continuous series. This is the second aspect of this constitutive paradox insofar as it takes the form of an indefinitely extensile series. The instant cannot include itself without differing from itself. With respect to the genesis of sense (Deleuze writes that the paradoxical element assures the bestowal of sense), this genetic or expressive aspect of the paradoxical element, Deleuze calls the paradox of indefinite proliferation in which an indefinite series is engendered when the

[58] Deleuze 1990, pp. 40–41.

[59] On the paradoxical element as quasi-cause see Deleuze 1990, p. 183, see also pp. 6–7, and 166.

[60] Deleuze 1994, p. 94, see also p. 1 ff.

sense of each name must be designated by another name of a higher order.[61] Moreover, in this respect, Deleuze quotes Husserl who writes that expression's "*productivity, its noematic production, is exhausted in the expressing* and with the *form of the conceptual* which is introduced with < the expression >." [emphasis in text][62] Intensity and its paradoxical structure engender the noema and its conceptual form, and in this way, intensity expresses intentionality as its actualized form.[63] With respect to the genesis of time, potential continuously comes to be actualized generating a more intense actualization, and if we follow the series regressively from more to less intense, then the zero degree of intensity exists only as a limit. Consequently, the genesis of time remains within immanence.

The indefinitely proliferating series only comes to be resolved insofar as the paradoxical element engenders the past and future series and secures their convergence in the caesura by making the two series continuously diverge.[64] This means that past and future continuously split off in two series from within the caesura, that neither the potentiality of the future nor the actuality of the past are already given except insofar as they diverge in the internal difference haunting the intensive series, and that the two series arise out of the caesura and do not precede it. This split constitutes the empty form of time as it arises in the third synthesis, and prepares the genesis of the other two syntheses. This is a temporal form of what, with respect to sense, Deleuze calls the paradox of sterile division. The indefinite proliferation of the intensive series comes to be fixed only when its genesis ceases and the series becomes actualized in a sterile double so as to form not a genetic instant of transformation but a past in itself, i.e. in moving from the third synthesis of time to the second. That the series comes to have been actualized halts the indefinite proliferation insofar as it is a genesis of the new, but in doing so it casts time across a new dimension, that of the past necessitating a different form of synthesis, the second synthesis of time.

The past in itself also embraces the totality of time, an empty form, and a series, albeit each of these has a different character than in the third synthesis. Paradoxically, the whole of time is also found in the second synthesis in which all that has been actual is bound together in a series continuously receding into the past, each phase of the past becoming more and more past. This second synthesis corresponds to what Husserl calls retention,[65] and each retentive phase retains prior phases including all that was retained in those prior phases and so on, forming a continuum of the

---

[61] Deleuze 1990, pp. 28–31. For the paradoxical element "assuring the bestowal of sense," see Deleuze 1990, pp. 51, 95 and passim.

[62] Husserl Ideas I, section 124, p. 296. Deleuze 1990, p. 32.

[63] Deleuze 1990, p. 299–300.

[64] Deleuze 1990, pp. 40, and 95.

[65] Contrary to what Kelly has written ("Husserl, Deleuzean Bergsonism and the Sense of the Past in General" Michael R. Kelly (2008), *Husserl Studies*, 24(1):15–30), Husserl's retention corresponds with Deleuze's second passive synthesis of time not with the first. Confusion seems to have arisen because of the use of the term 'retained' in Deleuze's discussion of the first synthesis of time (Deleuze 1994, pp. 70 ff.), but what is meant here is not retention in Husserl's sense. Deleuze, himself makes note of these different uses of "retention," see Deleuze 1994, p. 80.

whole of time insofar as it has become actualized. In the second synthesis of time the paradoxical element traverses the series of the past, but rather than begetting a genesis of something new, here the paradoxical element binds each continuously successive past to the still more past and continuously retains new actualities as they arise. These new actualities reverberate through the series of retentions as the paradoxical element traverses the series binding the new to the series as having actualized the protentions of some potentialities rather than others and thereby continuously, so that the totality of the past in itself is always an open totality.

The form of time in the continuously expanding totality of retentions is empty insofar as the retained past contains the future-present-past series as a singular form regardless of what concretely occurs as its content. Husserl attempts to illustrate this form in his famous time diagram.

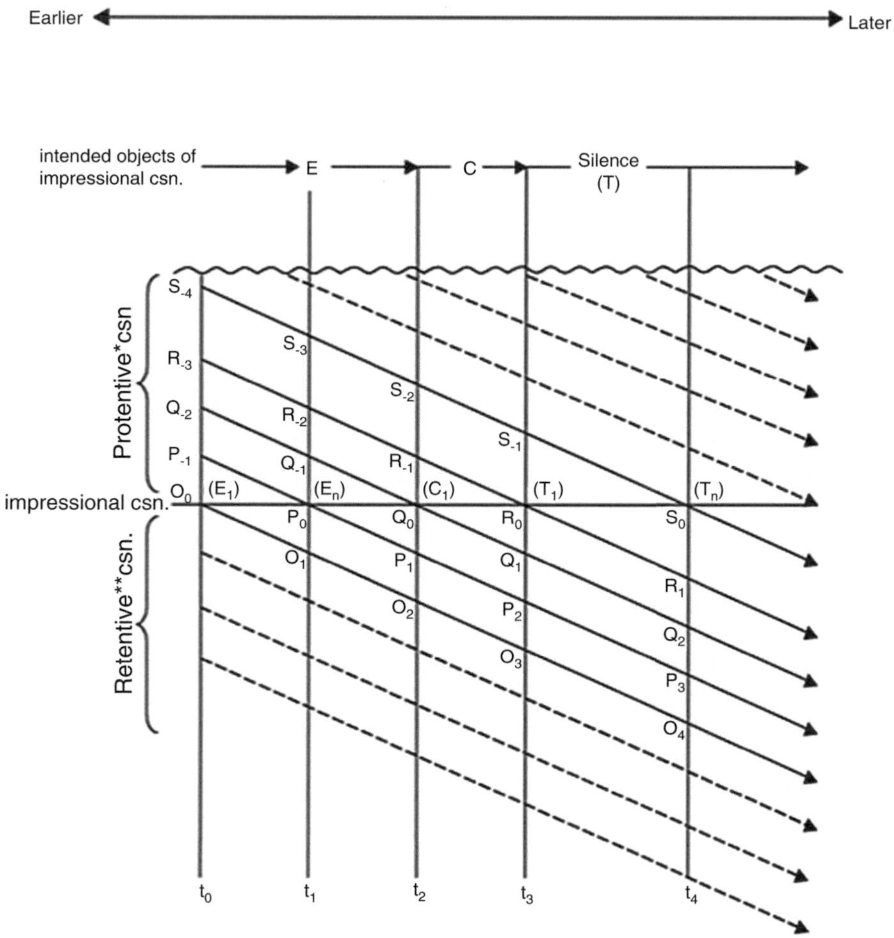

* with respect to the future but retentive with respect to earlier protendings
** with respect to the past but protentive with respect to later retendings

This more complete version of Husserl's time diagram is take from Robert Welsh Jordan, "Intentionality Notes" http://lamar.colostate.edu/~rwjordan/Notes/NotesInt.HTML.[66]

The form of time is that of the continuous future-present-past flow of protentions as they have come to be retained and continue to be retained either as having been or failing to have been actualized. Protentions are themselves retained as given within the series of retentions and while they are directed toward that which is to come, as the objects of protentions have come to be actualized, or as they fail to have come to be actualized so that that which had been projected onto the future has come to be cancelled out, those protentions with their objects or cancelled objects integrate continuously with already retained past phases. Phases, while distinguishable from one another are not discrete, and their continuity means that the phases of the temporal flow are not separable pieces, phases only form abstract parts of this continuity. Notwithstanding this continuity, the difference between the Event in the third synthesis, that which is not-yet, and the series of retentions in the second synthesis, that which has already been, is not abstract; the break between the two marks the difference between the two forms of synthesis.

In this second passive synthesis of time, the past in itself is constituted, on Deleuze's account, by four constitutive paradoxes, that the past is contemporaneous with the present it was, that the whole of the past coexists with the new present in relation to which it is now past, that the past pre-exists the passing present, i.e. that the past was never present nor was it formed after the present, and lastly that the whole past coexists with itself in varying degrees of tension and relaxation.[67]

That the past is contemporaneous with the present it was is a necessary condition for the present to pass since the present must pass simultaneously with it being present. Otherwise, no present could pass. It would be as if one present were simply replaced by another, as in the first passive synthesis of time, but the latter could never replace the former, for there then would be no continuity between the one present and the next; a new present would amount to a recreation of the whole of time, the second synthesis would collapse into the third and the present could never pass. Under such a condition of replacement, time would not be continuous and the present would have to leap over a gap in order to pass; the past could not be continuously retained – it could not even be formed. From Husserl's point of view, retention does not come after the present impressional moment; every present impressional moment arises contemporary with its being retained, every impression is retained at the same time as it occurs even though not every retained phase is impressional.[68] That the past in itself is founded on the present impressional moment is not a matter of temporal precedence; the form of time in the second synthesis, i.e. the protentional-

---

[66] Husserl's original truncated form of this diagram depicts only the series of retentions, see Husserl 1991 pp. 29 and 98, however it is clear from Husserl's expositions in these lectures and in later works that the diagram should be expanded as it is here.

[67] Deleuze 1994 pp. 81–84.

[68] Unfortunately, Husserl was not always as clear as he should have been on this point. See note 41 above.

retentional structure, belongs to every moment and is not acquired by present impressional moments after they occur – nothing can belong to a purely present impression which does not also belong to retention and to that structure.[69] The present impressional moment consists in an actual affection which founds this form of time and which necessarily shows itself in the running off of retentions. The phenomena of involuntary memory, famously exemplified by Proust's madeleine, shows that such affections, even though they are given passively, can be retained and recovered by active consciousness. Yet the affective phase can only be given insofar as it is abstracted from the concrete continuity of retentions in which it is always found synthesized with and inseparable from that continuity.[70]

One consequence of the contemporaneity of the past with the present that it was is that the past is never present since it has not been formed subsequent to the present. That which is grasped in the complex interconnected structure of retentions forms a synthesis of the whole of time, but a whole of time which was never present and which was not formed after the passing present. This positions the series of retentions as already there constituted through the second paradox that the whole of the past coexists with each new present as it becomes actualized and with respect to which the past has passed. The whole of the past is retained through the series of retentions and retentions of retentions in the interconnected structure of the whole of time. The past is thereby preserved in itself and is presupposed by the new present which arises out of the past.

That each new present presupposes the past in itself gives rise to the third paradox which constitutes the past as pre-existing the passing present. The past, already there, is presupposed by every passing present. That which has been retained in retentions and is a presentation of the past itself, is distinct from any form of representation of the past. However, the past is not merely a former present but is constituted by the bearing of the past on the passing present. The past is the synthesis of all of time in which present and future are dimensions. Since the structure of retentions always contains the form of the series of protentions, insofar as they have been retained, continuous with retentions holding onto all that has already been retained the whole line of passing presents is orthogonal to this continuous structure as is depicted by the vertical line in Husserl's time diagram. So the past does not belong to one present any more than another.

The second synthesis privileges the past so that the present is formed in relation to the complex structure of retentions, while the third synthesis privileges the future so that it continuously ungrounds time as belonging to the second synthesis. The past in itself retained in the second synthesis, however, bears on the metamorphosis engendered in the third synthesis insofar as actualization extends the lines of poten-

[69] This point contrasts Deleuze and Husserl sharply with Derrida for whom the split in time comes after an impossibly isolated present (Jacques Derrida, *Speech and Phenomena And Other Essays on Husserl's Theory of Signs*, translated by David B. Allison, Northwestern University Press, 1973, see pp. 48–69).

[70] On the present as an affection on which a phase of the retentional series is founded see Andrea 2010.

tialities which are selected from and have in the past already been projected onto the future. As bearing on the present the past in-itself structures the potentialities which can become actualized so that the present may pass, as Deleuze writes "[the past] is the in-itself of time as the final ground of the passage of time."[71] This is the fourth constitutive paradox, which brings together the other three; the paradox that the whole past co-exists with itself in degrees of tension and relaxation with the present as the most contracted degree of the totality of the past, i.e. "the new present… comes forth only by contracting this past"[72] Bergson's famous image of the past as a virtual cone attempts to depict these degrees of tension and relaxation, and marks not merely the bearing of the past on the present in the act of transformation, but also the selection of degrees of tension and release insofar as they come to be actualized.

Deleuze differs radically from Husserl in accounting for the genesis of representation, since possibilities (categories, genera, species, and even resemblances) themselves must have their genesis in the third synthesis, and it is in this sense that the third synthesis ungrounds the past. This differs from Husserl with respect to the way the past bears on the future; in addition and prior to protentions projecting already constituted possibilities emptily on the future. For Deleuze, virtual potentialities selected from the past in itself presage a multiplicity of lines of development embodied in distributions of singular points, of prominences, which may be prolonged in relation to one another along this multiplicity of lines. Prominences with their degrees of intensity are already retained insofar as they have arisen in the affections constituting the present impressional moment. Distributions of singular points are already retained in the complex interconnected structure of retentions in the form of distributions of prominences which exert different degrees of attraction in accord with their different degrees of intensity and which resonate with one another in relation to those degrees. These distributions and resonances, however, remain problematic with respect to which potential lines of development will have come to be actualized. Rather than presupposing already generated resemblances and possibilities, the past in itself bears, in the form of rhythmic distributions of aperiodic accented and unaccented beats, on the metamorphosis, the internal difference between potential and its actualization in the caesura. These distributions of singular and ordinary points, which are both engendered in the third synthesis and continuously return from the past in itself as it bears on metamorphosis in the caesura, organize and structure the potential multiplicity of lines of development which may become actualized.[73]

That the past in itself has been actualized does not imply that its virtual bearing on the present is actual. The virtual is real but not actual in the sense that the aperiodic rhythmic distribution of accents and their resonances in accord with their degrees of intensity are already implicit in the past in itself, while in accord with the

[71] Deleuze 1994 p. 82.

[72] Ibid.

[73] For more on the genetic force of these rhythmic structures see my "Rhythm: Assemblage and Event" *Strategies: Journal of Theory, Culture & Politics*, Vol. 15, Issue 1, 2002.

paradox of the pre-existence of the past are never present. The virtual contains concrete determinate structures but only in the form of problematic potentialities. These multiple structures of aperiodic rhythmic distributions of accents and their resonances effect "non-localizable connections, actions at a distance, systems of replay, resonance and echoes, objective chances…which transcend…temporal successions."[74] It is these structures which appear as actualized in the present, but the present is never fully actualized in accord with the third synthesis and is only the most contracted, i.e. intense, form of the past in itself which is past and future, retentional and protentional, but was never present in accord with the second synthesis. So, for there to be an actual present, a present that passes due to the insistence of the virtual in the second synthesis, requires another synthesis, the first, a passive synthesis of the living present. The actual present measures the temporal realization of the event by bringing a sequence of presents under the form of the same covering over difference in itself, acting in the third and second syntheses, with resemblances constituting contents which supervene upon these virtual structures.

Both Husserl and Deleuze discuss the first passive synthesis of time in the form of the notion[75] of a living present and a *habitus*; the former because it is in the first synthesis that time is experienced as going from one present to another, and the latter because it is within the living present that contingent habitual connections are made between temporal objects. Here we return to Hume and Hume's empiricism. Hume claims that time is merely the succession of perceptions in the mind,[76] and that this psychological succession of perceptions becomes ordered by the laws of association. This account tends towards circularity because the succession here must be an immanent and temporal succession, since it would not be enough to account for temporal passage were it a McTaggart C-series.[77] Hume also insisted, in accord with the sequence of presents combined in the first synthesis, that this succession consisted of indivisible moments so that the passage of the present is utterly lost.[78] Neither Husserl nor Deleuze fall into these dilemmas since they can appeal to the complex structure of retention, in the second synthesis, as constituting time prior to and as a condition of continuous temporal passage and thereby of succession in the living present. However, Husserl was never clear regarding the nature of the living present, which he sometimes mistakenly describes as the emptying of retentions

---

[74] Deleuze 1994 p. 83.

[75] This notion has a dubious ancestry which unfortunately calls up the notion of the specious present or "saddle-back" posited by William James. Neither Husserl nor Deleuze accept James' claim that "the practically cognized present is no knife-edge, but a saddle-back, with a certain breadth of its own on which we sit perched…" (William James, *The Principles of Psychology*, Vol. 1, Dover Publications, 1950, p. 609). So, insofar as Husserl, although he is not always clear about this (see notes 68 and 41) and Deleuze write of "the living present" they are referring to the experience and connection by association of successive presents.

[76] David Hume, *Treatise of Human Nature*, edited L. A. Selby-Biggs, The Clarendon Press, Oxford, 1960, pp. 34–35.

[77] John McTaggart Ellis McTaggart (1908) "The Unreality of Time," *Mind*, New Series, no. 68, pp. 457–474.

[78] Hume 1960, p. 31.

of their intuitive content[79] and sometimes in terms of degrees of increasing obscurity with respect to the past. The former, however, cannot occur since retentions must always be intuitively filled, as Husserl usually recognizes and which accords with Hume who rightly says that time cannot be separated from its contents. But even the latter approach requires clarification, since it is not clear that obscurity of the contents of retentions do not generally become past to the same degree or necessarily in a continuous manner.

Deleuze takes the first synthesis of time, not primarily as a synthesis of contents within time, although he accepts the Humean stricture on the inseparability of time and its contents, but as a synthesis of time, as a synthesis connecting temporal moments into a living present. These connections order a series of presents which pass and which form a series of past presents[80] connected to a more recent present and ultimately to the present present, through a double series of associations. The series of presents as such is formed in the first synthesis by holding former presents together with later presents, by bridging the gulf between two successive independent presents, by extracting the difference between successive presents as passively contemplated by a noesis which synthesizes the two presents as similar instances in an asymmetric order, initially as present present and former present. This similarity arises through contemplating[81] these different presents as the same insofar as they effect a noesis which extracts their difference. Deleuze references Hume's thesis that "Repetition [of the present] changes nothing in the object repeated, but does change something in the mind which contemplates it."[82] Of course, the passage from one present to the next could not take place in this first synthesis alone[83] which is why the first synthesis must be conditioned by the second and the third. For, since the present passes continuously it could not consist, as Hume claims,[84] merely in a sequence of indivisible presents. The sequence of presents erect multiple contemplations, which act between the presents combined; these contemplations still belong to passivity and do not form the unity of an active ego. What they form instead is a *habitus*, a foundation upon which habits of association can be formed, a passively formed sequence of presents upon which a secondary passivity of associative connections may be formed. Such habits connect the resemblances and contrasts formed in the linked sequence of presents at a secondary level in accord with their contents through what Husserl calls distinguishing and identifying syntheses.

[79] For a defense of this view see Nicolas de Warren, *Husserl and the Promise of Time: Subjectivity in Transcendental Phenomenology*, Cambridge University Press, 2009, p. 281 ff. and passim.

[80] These are not the same as retentions in the sense of the second synthesis, despite Deleuze's use of the term, see note 65.

[81] Deleuze 1994, p. 74 see also p. 78 for different expressions of this synthesis. It should also be noted that the text uses the term contraction for the fusion of successive presents. We shall avoid this term since what is at issue differs radically from the sort of contraction evinced in the second synthesis, see Deleuze 1994, p. 82 ff.

[82] Deleuze 1994, p. 70.

[83] Ibid., pp. 272–277.

[84] Hume 1960 p. 31

It is through the careful study of these that Husserl is able to ground judgments in aesthesis.[85] These passively acquired habits lay the foundation for the active synthesis of memory and of the ego as unified and unifying and which can then represent the present and former presents in an extra dimension through which reflection is possible.[86]

References

1976. Die Krisis der europäischen Wissenschaften und die transzendentale Phänomenologie. In *Eine Einleitung in die phänomenologische Philosophie*, ed. Walter Biemel. The Hague: Martinus Nijhoff.

Andrea, Staiti. 2010. The primacy of the present: Metaphysical ballast or phenomenological finding? *Research in Phenomenology* 40(1): 34–54.

Aristotle. 1941. Metaphysics. In *The basic works of Aristotle*, ed. Richard McKeon, 689–934. New York: Random House.

Cooper, Grosvenor, and Leonard B. Meyer. 1960. *The rhythmic structure of music*. Chicago: The University of Chicago Press.

Crowell, Steven Galt. 2001. *Husserl, Heidegger, and the space of meaning: Paths toward transcendental philosophy*. Evanston: Northwestern University Press.

de Warren, Nicolas. 2009. *Husserl and the promise of time: Subjectivity in transcendental phenomenology*. Cambridge: Cambridge University Press.

Delanda, Manuel. 2002. *Intensive science and virtual philosophy*. London/New York: Continuum International Publishing Group.

Deleuze, Gilles. 1980. Interview with Catherine Clement. *L'Arc* (revised edition) 49: 94–99.

Deleuze, Gilles. 1983. *Nietzsche and philosophy*. New York: Columbia University Press.

Deleuze, Gilles. 1991. *Empiricism and Subjectivity: An Essay on Hume's Theory of Nature*. Trans. Constantin V. Boundas. New York: Columbia University Press.

Deleuze, Gilles. 1990. *The logic of sense*. London/New York: The Athlone Press.

Deleuze, Gilles. 1994. *Difference and repetition*. New York: Columbia University Press.

Deleuze, Gilles. 2001. Immanence: A life. In *Pure immanence*. New York: Zone Books.

Deleuze, Gilles, and Felix Guattari. 1986. *Anti-Oedipus: Capitalism and Schizophrenia*. Trans. Robert Hurley et al. Minneapolis: University of Minnesota Press.

Deleuze, Gilles., and Felix Guattari. 1987. *A Thousand Plateaus: Capitalism and Schizophrenia*. Trans. Brian Massumi. Minneapolis: University of Minnesota Press.

Deleuze, Gilles, and Felix Guattari. 1994. *What is philosophy?* New York: Columbia University Press.

Derrida, Jacques. 1973. *Speech and Phenomena and Other Essays on Husserl's Theory of Signs*. Trans. David B. Allison. Evanston: Northwestern University Press.

Descartes, René. 1984. Meditations on first philosophy. In *The philosophical writings of descartes*, vol. 2. Trans. John Cottingham, et al. Cambridge: Cambridge University Press.

Hughes, Joe. 2008. *Deleuze and the genesis of representation*. London/New York: Continuum.

Hume, David. 1960. Book I, Part II Of the Ideas of Space and Time. In *Treatise of human nature*, ed. L.A. Selby-Biggs. Oxford: The Clarendon Press.

Husserl, Edmund. 1939. *Erfahrung und Urteil*, Untersuchungen zur Genealogie der Logik. Redigiert und heruasgegeben von Ludwig Landgrebe. London: Allen and Unwin.

[85] This is the project of Husserl 1939.

[86] Deleuze 1994, pp. 103–104 ff.

Husserl, Edmund. 1969. *Formal and Transcendental Logic*. Trans. Dorion Cairns. The Hauge: Martinus Nijhoff.
Husserl, Edmund. 1976. *Die Krisis der europäischen Wissenschaften und die tranzendentale Phänomenologie*. The Hague: Martinus Nijhoff. English edition: Husserl, Edmund. 1970. *The Crisis of European Sciences and Transcendental Phenomenology* (trans: Carr, David). Evanston: Northwestern University Press.
Husserl, Edmund. 1981. Husserl's Inaugural Lecture at Freiburg im Breisgau (1917). Trans. Robert Welsh Jordan. In *Husserl: Shorter Works*, ed. Peter McCormick and Frederick Elliston. Notre Dame: University of Notre Dame Press.
Husserl, Edmund. 1982. *Cartesian Meditations: An Introduction to Phenomenology.* Trans. Dorion Cairns. The Hauge: Martinus Nijhoff Publishers.
Husserl, Edmund. 1983. *Ideas Pertaining to a Pure Phenomenology and to a Phenomenological Philosophy.* Trans. F. Kersten. The Hauge: Martinus Nijhoff Publishers.
Husserl, Edmund. 1991. *On the Phenomenology of the Consciousness of Internal Time (1893–1917).* Trans. John Barnett Brough. Dordrecht: Kluwer Academic Press.
Husserl, Edmund. 2001a. *Logical Investigations*, vol. 1. Trans. J.N. Findlay. New York: Routledge.
Husserl, Edmund. 2001b. *Analyses Concerning Passive and Active Synthesis: Lectures on Transcendental Logic.* Trans. J. Anthony. Steinbock: Kluwer Academic Publishers.
Husserl, Edmund. 2008. *Introduction to Logic and Theory of Knowledge: Lectures 1906/07.* Trans. Claire Ortiz Hill. Dordrecht: Springer.
James, William. 1950. *The principles of psychology*, vol. 1. New York: Dover Publications.
Jordan, Robert Welsh. 2008a. Phenomenologically related work of Robert Welsh Jordan. http://lamar.colostate.edu/~rwjordan/Notes/W-NotesHUS.HTML#E_time,_as_the_Universal_form. Accessed June–Aug 2012.
Jordan, Robert Welsh. 2008b. Phenomenologically related work of Robert Welsh Jordan. *Notes on the Intentionality of Consciousness.* http://lamar.colostate.edu/~rwjordan/Notes/NotesInt.HTML. Accessed June–Aug 2012.
Jordan, Robert Welsh. 2008c. Phenomenologically related work of Robert Welsh Jordan. http://lamar.colostate.edu/~rwjordan/Notes/W-NotesHUS.HTML#E_objects_[Gegenst%E4nde]_and_objectivating_[Vorstellen]. Accessed June–Aug 2012.
Jordan, Robert Welsh. 2008d. Phenomenologically related work of Robert Welsh Jordan. http://lamar.colostate.edu/~rwjordan/Notes/W-NotesHUS.HTML#E_formal_essences,_superordinate_to_material_essences. Accessed June–Aug 2012.
Kelly, Michael R. 2008. Husserl, Deleuzean Bergsonism and the sense of the past in general. *Husserl Studies* 24(1): 15–30.
McTaggart, J.M.E. 1908. The unreality of time. *Mind, New Series* 17(68): 457–474.
Murphy, Richard T. 1980. *Hume and Husserl: Towards radical subjectivism.* The Hague: Martinus Nijhoff Publishers.
Nietzsche, Friedrich. 1968. *The Will to Power.* Trans. Walter Kaufmann. New York: Vintage Books.
Nietzsche, Friedrich. 1974. *The Gay Science*. Trans. Walter Kaufmann. New York: Vintage Books.
Nietzsche, Friedrich. 2006. *Thus Spoke Zarathustra*. Trans. Adrian Del Caro. Cambridge: Cambridge University Press.
Russell, Bertrand. 1903. *The principles of mathematics*. Cambridge: Cambridge University Press.
Smith, Daniel W. 2001. The doctrine of univocity: Deleuze's ontology of immanence. In *Deleuze and religion*, ed. Mary Bryden. London: Routledge.
Turetzky, Philip. 1998. *Time*. London/New York: Routledge.
Turetzky, Philip. 2002. Rhythm: Assemblage and event. *Strategies: Journal of Theory Culture & Politics* 15(1): 121–138.
Williams, Bernard. 2006. What might philosophy become? In *Philosophy as a humanistic discipline*. Princeton/Oxford: Princeton University Press.

Chapter 12
Change's Order: On Deleuze's Notion of Time

Dror Yinon

Abstract In this paper I focus on the second chapter of Deleuze's *Difference and Repetition* in order to point out the distinctive traits of his theory of time and in particular to clarify Deleuze's claim that change is the form of time. I argue first, that Deleuze's theory of time is related to the transcendental tradition in which the discussion of time is inseparable from the understanding of subjectivity. Second, that his notion of time is distinct with regard to this tradition because Deleuze's analysis of passive subjectivity reveals that time, though not objective and detached from subjectivity, is not attributable to it; it is rather subjectivity (or the different layers of subjectivity) that is generated through the syntheses of time. In order to show this I discuss Deleuze's three syntheses of time focusing on the character of the time that is fundamental in each synthesis (present, past, future). Finally, I offer a clarification of Deleuze's notion of change as the form of time through a discussion of McTaggart's argument for the unreality of time and the understanding of change that it presupposes.

Keywords Deleuze • McTaggart • Time • Subjectivity • Repetition • Ideality • Change • Transcendental philosophy

12.1 Introduction

One of the most basic distinctions within the philosophical discussion of time is the one between 'objective', 'natural' or 'physical' time on the one hand and 'subjective', 'phenomenological' or 'human' time on the other. This branching of the discussion of time is supposed to express intuitively separated interests regarding time: the first is an interest in the nature of time while the second is an interest in the experience of time, or, lived-time.

D. Yinon (✉)
The Program for Hermeneutics and Cultural Studies, Bar-Ilan University, Ramat Gan 5290002, Israel
e-mail: dror_yinon@yahoo.com

Y. Dolev, M. Roubach (eds.), *Cosmological and Psychological Time*, Boston Studies in the Philosophy and History of Science 285,
DOI 10.1007/978-3-319-22590-6_12

This distinction has become pretty much entrenched in the constantly growing universe of literature on time and reflects a peaceful division of labor where each discussion is generally minding its own business. However, one of the central currents of modern philosophy, transcendental philosophy, suggests a project that challenges this very separation through an attempt to ground what we call 'objective time' in the subject, thus making the nature of time a primary concern for any investigation concerning the nature of subjectivity.[1] This might complicate things as it brings to the combination of the terms 'subject' and 'time' other senses that are not intended in the 'subjective time' of our initial distinction, but even this point alone only makes it more interesting suggesting that the discussion of time might be missing something.

The extent to which the disregard of this more complicated possibility might be constitutive of the distinction itself can be learned, for example, from the fact that although Kant attempted to attach objective time to subjectivity, McTaggart (to whom we will return later) understood Kant as simply rejecting the reality of time. For McTaggart time would be real if it would be possible to predicate temporal properties of events[2]; yet for Kant "time is not an empirical concept that is somehow drawn from an experience,"[3] but rather grounds a priori anything that can be given within experience. Thus, time's being an a priori condition for experience is characteristic of the transcendental tradition. This means that the question is not whether time is real, as time is one of the conditions that set the terms of what it is to be real, but rather how time is to be understood once we accept its transcendental status.

The both attractive and controversial feature of the transcendental treatment of time is the insistence upon a deep and intimate relation of time to subjectivity which can be explicitly formulated as dependency relations, as Kant had put it:

"…if we remove our own subject or even only the subjective constitution of the senses in general, then all constitution, all relations of objects in space and time,

[1] In his account of the history of temporality which mostly bears on thinkers related to the transcendental tradition, David Couzens Hoy distinguishes between human time, or, temporality and universal time, or simply, time and restricts his discussion to the former. Couzens Hoy argues that ignoring the distinction leads to philosophical labyrinths, thus he offers to see the problem of the "time of our lives" as separate from the problem of the "time of the universe", see Couzens Hoy (2009, pp. xii–xvii). The basic assumption of the position I present here, with regard to the transcendental tradition is different. Yet, although I believe this characterization is true of philosophers related to the transcendental tradition, Husserl is an exception, at least in his lectures on internal time consciousness. In these lectures he explicitly suspends the question of the relation of the experience of time (subjective time) to real time (objective time), see Husserl (1991, §1), but compare Michalski (1997, pp. 128–141) who rejects such a distinction and takes Husserl's phenomenology of subjective time as an analysis of the conditions of the possibility of objective time. Nevertheless, Husserl is no exception with regard to the intimate relation of time to subjectivity. For an analysis of the centrality of this relation in Husserl's thought see De Warren (2009).

[2] See McTaggart (1908). I will discuss Deleuze's notion of time in relation to McTaggart's assumptions in the final section of this paper.

[3] Kant (1998, B46).

indeed space and time themselves would disappear, and as appearances they cannot exist in themselves, but only in us."[4]

While time is grounded in subjectivity, subjectivity should not be thought of as independent of time: it is clear, within the transcendental tradition, that at least a significant part of an account of subjectivity must be given in temporal terms and this direction was developed by successors of the transcendental tradition such as Heidegger or Merleau-Ponty who writes: "to analyze time is not to follow out the consequences of a pre-established conception of subjectivity, it is to gain access, through time, to its concrete structure."[5] Clearly this direction of development only strengthens the reason for holding the above Kantian contention that had there been no subject, time would disappear as well.

Yet this does not mean that the transcendental position gives up its ambition to account for objective time. The idea is that even if time, after being investigated transcendentally is dependent upon the subject, it should empirically or in the mode of everydayness or from a natural attitude be grasped *as if* it were independent of the subjectivity in which it is grounded. Thus it seems that the transcendental position attempts to accommodate two somewhat opposing intuitions: the first is that time must be comprehensive with regard to everything that happens and as such it should be independent of anything that occurs in time while the second is that much of the sense that we relate to time is possible only with regard to a subject for which it is meaningful. This is echoed in the manner in which Merleau-Ponty articulated the task of his phenomenology of perception: "We cannot remain in this dilemma of having to fail to understand either the subject or the object. We must discover the origin of the object at the very centre of our experience; we must describe the emergence of being and we must understand how, paradoxically, there is *for us* an *in-itself*'.[6] This general articulation of the task holds particularly in the case of time where Merleau-Ponty writes that "time presupposes a view of time… [it] is not a real process, not an actual succession that I am content to record. It arises from *my* relation to things", while on the other hand "the objective world is too much of a plenum for there to be time".[7] Transcendental philosophers have tried to solve this in-built tension by securing the continuity between the objective notion of time and its subjective ground. Thus Kant secured the passage from the subjective origin to the objective use of time by arguing that time is infinite and is given as an unlimited representation, while Heidegger insisted that while temporality is primordially

[4] Kant (1998, B59). Recently, Quentin Meillassoux presented a critique of this transcendental tradition from within the French philosophical context (he names the position he criticizes 'correlationism' in order to include philosophers which are not part of the transcendental tradition, but it is this tradition that motivates his critique and it remains his main target). Indeed, the problem of the relation of the subject or thought to time considering the fact that time is prior to the existence of any subject is the ground upon which he bases his speculative realism. See Meillassoux (2008, Chap. 1).

[5] Merleau-Ponty (2002, p. 477).

[6] Merleau-Ponty (2002, pp. 82–83).

[7] Merleau-Ponty (2002, pp. 477–478).

finite it provides the meaning structure that confers sense upon 'objective' conceptions of time.[8]

Gilles Deleuze's *Difference and Repetition* should be read – and the case of his discussion of time makes no exception – as a part of this transcendental tradition which sets its framework and provides it with its problems even if Deleuze is deeply invested in the thought of other philosophers such as Bergson and Nietzsche. Deleuze seems to agree with the transcendental philosopher's basic claim yet he affirms it with a qualification that I believe is crucial, when he writes: "time is subjective, but in relation to the subjectivity of a passive subject."[9] This comment gives us an access to a view that challenges not only the basic objective/subjective distinction but also the transcendental grounding of time in subjectivity.

Before I turn to Deleuze's thought on time it is worthwhile to list again the main features of a transcendental approach to time. According to such an approach time cannot be derived from experience but should be taken as an a priori condition of experience. Time is dependent upon subjectivity; subjectivity is the origin of time which in turn provides the meaningful content of subjectivity itself. Finally, it is committed to establish a connection between the subjective provenance of time and the intuitive, though misleading, apprehension of time as independent of the subject such that the subjective origin will contribute to the explanation of the fact that the ordinary and scientific attitudes towards time are so distanced and even opposed to it.

Now, it is with no doubt that Heidegger and Merleau-Ponty could accept a certain version of Deleuze's comment that the subjectivity to which time is attached is passive. But what I would like to argue here is that Deleuze's comment is more subversive than it appears: the investigation of time with relation to a passive subjectivity leads to the claim that time transcends subjectivity, or that subjectivity is not the origin of time. This does not mean, however, that we abandon the search of time through subjectivity and settle for an objective time but rather that the transcendental inquiry itself reveals different layers of subjectivity that are generated by three syntheses of time. Thus, following Deleuze, we might want to revise our initial subjective/objective distinction. As we already saw the transcendental approach holds a richer sense of the subjective than the epiphenomenal human experience implied in the initial distinction; yet it still retains the distinction itself only to suggest a hierarchy in which objective time is grounded in the time originating from transcendental subjectivity. Deleuze's account of time goes further to question the distinction itself: as we shall see, the analysis of time goes through the analysis of subjectivity but it reveals that time is not dependent upon the subject; and yet, time's transcendental nature makes it understandable only with relation to the subjectivity that is generated by it.

The transcendental investigation of subjectivity remains then, the access to an understanding of time, and time is still regarded as a condition of experience; however, time is not dependent upon subjectivity in the sense that subjectivity is revealed

---

[8] See Kant (1998, B48) and Heidegger (1962, pp. 464–480).

[9] Deleuze (1994, p. 71).

as its origin. The important consequence of Deleuze's investigation of subjectivity is that time is ideal but not subjective, that is, Deleuze suggests a notion of ideality which is not attributed to subjectivity. In other words, Deleuze rejects the too immediate connection between ideality and interiority; indeed, repetition (together with difference) is supposed to embody such an ideality. This might accommodate better the two intuitions which I mentioned earlier: the intuition that time is not dependent upon a subject and the one that sees time as ideal. Thus, what might be wrong in the transcendental approach is the reduction of ideality to subjectivity.

In what follows I will present briefly Deleuze's central discussion of time in the second chapter of *Difference and Repetition* and then point to some of the consequences of this Deleuzian notion of time, with regard to the transcendental approach to time on the one hand and on the other, with regard to McTaggart's analysis of time.

12.2 Deleuze's Syntheses of Time

The context of Deleuze's discussion of time requires a brief clarification of his project in *Difference and Repetition*. Deleuze's project in *Difference and Repetition* is to give an alternative to the Kantian account of the transcendental field and its relation to experience.[10] Without going into the details, Deleuze argues that the Kantian notion of the transcendental is limited to conditions of experience that only account for the possibility of something to appear or to be given while what a transcendental account should do is reveal the principles that are active in the production of experience and its subject. Thus, rather than providing an analysis of fixed constraints on experience as Kant does, Deleuze attempts to find active principles that will account for the dynamic of experience while not a priori fixing its limits. This of course, is not merely a matter of philosophical taste but rather emerges from a critique of the Kantian account of the relations between faculties and the structure of the transcendental based upon it. Finally, this critique leads Deleuze to argue that we need to understand anew the notions of difference and repetition. These notions, Deleuze claims, play an important though an implicit role within the Kantian accepted framework of experience. A crucial point for Deleuze's argument is that difference and repetition are transcendentally active principles which do not presuppose subjectivity – a cognitive representation-based agency that constitute objects and has an experience-independent status – but rather explain genetically its production within experience. The context, then, of Deleuze's discussion of time is his investigation of this new notion of repetition.[11]

These considerations lead us back to Deleuze's comment with regard to passivity. Starting with repetition as a principle, whatever is generated is its product. Thus

---

[10] For a recent account of Deleuze's relation to Kant's thought see Lord (2012).

[11] For more comprehensive discussions of Deleuze's philosophy of time see Turetzky (1998, pp. 211–229) and Williams (2011).

if subjectivity is generated through repetition, whatever happens, happens to the subject and not by the subject. But, it would be too hasty to identify subjectivity with everything that is formed through the activity of repetition, even if it necessarily has an ideal aspect to it. Now, repetition is the transcendental principle that generates subjectivity and also allows us to see its limits. It does so with regard to both of its aspects: its temporal character and its psychological structure (which will not be discussed here). As we already saw the view of subjectivity as the origin of time is deeply rooted in the transcendental tradition, and thus Deleuze sets an alternative to this basic transcendental convention by showing that there is indeed a connection between subjectivity and temporality but this connection is established by repetition that produces different syntheses of time which reveal the different layers of subjectivity or better, ideality (within which subjectivity also emerges).[12]

The basic insight with which Deleuze starts his discussion of repetition and time is that "repetition changes nothing in the object repeated, but does change something in the mind which contemplates it".[13] Thus all that is needed for repetition is a certain kind of ideality, in this case a mind in which contemplation of repetition occurs, and hence a synthesis is generated. This synthesis is passive, meaning that on the one hand it is constitutive of time, but it is passive as "it is not carried out by the mind, but occurs *in* the mind which contemplates prior to all memory and reflection."[14] This synthesis, then, does not presuppose active subjectivity – as we shall see a contemplating mind for Deleuze is completely different from the representation that active subjectivity requires – and the synthesis that is being generated is temporal. As Deleuze's discussion of time is at the same time a discussion of subjectivity, the different temporal syntheses are attached to different faculties that unfold passivity on all its facets and by way of doing that suggest a fresh look at the nature of time. Deleuze's syntheses proceed, then, through the different temporal dimensions and at the same time through the different phases of passive subjectivity: with regard to the order of passivity the syntheses proceed from imagination and sensibility through memory to thought, and with regard to temporal dimensions it starts with the present continues through the past and culminates with the future.

The first, most simple, activity of repetition is the contraction of instants. Deleuze borrows his examples here from Hume and Bergson.[15] Take the Bergsonian example

[12]This is also expressed in the title of the chapter of *Difference and Repetition* dealing with time which is "Repetition for Itself". As Deleuze writes: "The constitution of repetition already implies three instances: the in-itself which causes it to disappear as it appears, leaving it unthinkable; the for-itself of the passive synthesis; and, grounded upon the latter, the reflected representation 'for-us' in the active syntheses", (Deleuze 1994, p. 71). Repetition for itself is ideal: it happens in a contemplating mind but it is not interior to it, it does not come of the subject. Repetition conceived as the act of the subject comes only with the 'for us' of representation and is actually grounded in the ideality of the repetition that is attached to passivity.

[13]Deleuze (1994, p. 88).

[14]Deleuze (1994, p. 71), italics in the original.

[15]The differences between the Humean and Bergsonian cases are not important for me here, for Deleuze's discussion of them see, Deleuze (1994, pp. 71–72). It is interesting to note that although Deleuze wrote books both on Hume and Bergson in the years preceding *Difference and Repetition*,

of the church bell announcing the hour, let's say, 4 o'clock. Now, what happens here is that the ringing of the bell repeats, but the repetition should be in something else that contemplates the ringing of the bell otherwise there would be no second, third and fourth ringing. Indeed, what happens here is that we get repetition through what Deleuze calls contraction. The impression of 4 o'clock is synthesized by contracting certain instants, in this case, the ringing of the bell. This repetition through contraction is an ideal change as it cannot be attributed to the thing, but it does not originate from the subject either. The level where such changes take place according to Deleuze does not involve representation and memory which are constitutive of subjectivity in its active version; rather, it belongs to what Deleuze, following Hume, calls imagination.[16] The ringing of the bell immediately generates a contraction that happens to an individual to whom the execution of such a synthesis cannot be attributed. Now, this example is only an illustration and already belongs to a much higher level than the one at which Deleuze aims. What Deleuze points to is an organic level that precedes sensibility and actually forms it. Thus, a more adequate example that Deleuze brings is that of heart beats. The heart beats, that is, it contracts and then dilates and then contracts again etc. This beating of the heart is repetitive and as such involves what Deleuze calls 'contemplation'. This contemplation does not represent the instants; it synthesizes a beat out of them (thus the contraction that constitutes this level of repetition includes in the case of a heartbeat the dilation too). Now, such syntheses on the organic level occur incessantly in every organism and thus generate them as habit-constituted organisms. Forming habits, then, is a basic expression of repetition as contemplation. This means that ideality already appears on the basic organic level. Indeed Deleuze writes: "a soul must be attributed to the heart, to the muscles, nerves and cells, but a contemplative soul whose entire function is to contract a habit. This is no mystical or barbarous hypothesis. On the contrary, habit here manifests its full generality: it concerns not only the sensory motor

---

and at least in the case of Bergson, time is a central issue for the whole of his philosophy, these examples were not discussed in those studies. This is especially interesting in the case of Bergson where Deleuze's discussion is even opposed to the tenets of Bergson's philosophy of time, even if it borrows examples from it and uses it extensively in the second synthesis. This means that Deleuze's thought of time is not continuous with his studies of other philosophers and that *Difference and Repetition* suggests something new compared to previous studies.

[16] See Deleuze (1994, pp. 71, 76). It should be noted that in Deleuze's study of Hume, Deleuze already links the thought of Bergson to Hume's claiming that "it is not necessary to force the texts in order to find in the habit-anticipation most of the characteristics of the Bergsonian *Durée* or memory" (Deleuze 1991b, p. 92). However, the role played by the imagination and the conception of time developed there is very different from Deleuze's own (for Deleuze's discussion of Hume's synthesis of time see Deleuze 1991b, pp. 92–96). The same goes for Deleuze's use of Bergson in the context of the imagination which is opposed to the Bergsonian focus (central also in Deleuze's study of his philosophy) on memory and consciousness as key elements in the understanding of duration. Moreover, in the study of Bergson the affinity between Hume's habit and Bergson's duration is not mentioned. See Deleuze (1991a, pp. 51–72).

habits that we have (psychologically), but also, before these, the primary habits that we are; the thousands of passive syntheses of which we are organically composed".[17]

This synthesis is the synthesis of the living present. It is finite not because we cannot think of an infinite present but rather because an organism has limits to its contraction capacity. This gives the organism a specific duration of present, a specific rhythm: "[contraction] necessarily forms a present which may be exhausted and which passes, a present of a certain duration which varies according to the species, the individuals, the organisms and the parts of organisms under consideration".[18] Within this living present, past and future do not appear as independent moments but rather as dimensions of the present itself and their sense is organic: within the living present, past is heredity and future is anticipation (Deleuze's example is a chicken that pecks a grain while already anticipates the next).

This first synthesis leads to an active synthesis that involves the understanding and memory and it is with this active synthesis that we get our ordinary and intuitive notion of time where the past and the future are thought of as moments that are separated from the present. In distinction from the former synthesis which is the rhythm of the basic levels of individual entities and hence its span is limited, this synthesis is an intellectually based synthesis which nothing prevents from conceiving time as infinite, or at least, indefinite. An essential feature of this synthesis is representation's role in it: in temporal terms, representation means that the actual present represents a former present. Thus, the distinctive trait in this synthesis is that when a former present is represented, as in an instance of what Deleuze terms 'active memory', it is represented in such a way that the actuality of the representing present is also being reflected. What is traditionally conceived of as subjectivity is related, according to Deleuze, to this active synthesis. It is from the standpoint of this synthesis and the notion of time projected by it that time can be divided between its independent cosmological objective sense and its subjective experience. However, it is the inquiry into the passive synthesis of time that enables us going beyond the distinction between objective and subjective time. We shall move on then on the passive syntheses path.

The first synthesis presupposes that the present can pass. This means that another time, or more precisely, another synthesis of time must take place in order to allow for the passing of the present; the living present cannot provide the grounds for its own movement: "This is the paradox of the present: to constitute time while passing in the time constituted. We cannot avoid the necessary conclusion – *that there must be another time in which the first synthesis of time can occur*".[19] In this second synthesis which is of the past, the conception of memory as a function of reproduction

[17] Deleuze (1994, p. 74). Recently Elizabeth Grosz claimed that for Deleuze habit is "the condition for the emergence of time itself" (Grosz 2013, p. 231). This is misleading since for Deleuze time involves a complex relation between several syntheses. Habit plays a role only in the living present but not in the pure past or the future; thus it cannot be taken as a condition for the emergence of time itself.

[18] Deleuze (1994, p. 77).

[19] Deleuze (1994, p. 79), italics in the original.

of what formerly was present, as it is conceived in the active synthesis, is denied; rather, it is a past that precedes the presents and providing them the ground upon which they can pass: "habit is the originary synthesis of time, which constitutes the life of the passing present; memory is the fundamental synthesis of time which constitutes the being of the past (that which makes the present pass)".[20]

Deleuze's discussion of this notion of the past is rich and complicated but I wish to focus on the following insight: the past of the second synthesis is a past that was never present because as we just saw, in order to pass, the present must always already presuppose a past; thus, this past is different from a present that had passed and became a former present. This past, then, precedes the present and functions as its a priori condition while in terms of the faculties involved, transcendental memory is presupposed by the living present, hence, imagination and sensibility, and the same is true with regard to the understanding and memory of the active version of the second synthesis. Finally, this past contains all presents, actual or former, and thus it functions like the global version of the act of contraction in the first synthesis: in the first we had a contraction that captured a finite set of instants to form a habit, a rhythm and a local present, and with the second we get a past that contracts all presents (if the passing of the present presupposes such a pure past, it must contract all presents because no present can be thought of as outside it). Thus, the succession of presents that occur on the background of a pure a priori past is gained, but this synthesis also makes time closed upon itself, creating a circle where former presents and actual presents co-exist. There is, evidently, much to be said of this situation but what is important for us here is that by providing the ground where the present can pass, by allowing movement, time becomes attached to events, the passing presents.

The relation of time to events as expressed in the second synthesis is precisely what calls for a third synthesis. The pure past contains every possible event but this complete containment makes the movement within the system of the past a seeming movement which makes time an epiphenomenon of a fixed space of possibilities. Time turns out to be the measure of events, not their condition. Movement becomes containment that falsifies movement itself, hence, time. Thus, time does not complete its structure with movement, just as subjectivity does not reach its highest point in memory. In order to open the closed system of the pure past it must be traversed by one last synthesis, that of the future.

As repetition acts on all levels of subjectivity (thus repeating differentially its own act) it acts also within thought. This aspect of the third synthesis is crucial for my argument that it is through the investigation of subjectivity that we reach beyond the subjective/objective distinction with regard to time. Deleuze uses here Kant as claiming that the act that determines the self, the "I think", is realized only through the form of time in which the existence of the subject can be given. This immediately poses the problem of the determination of the subject as executing this act of determination. This is impossible since in order to do that we would have to have another time in which the subject in its act of determination is given. Facing this

[20] Deleuze (1994, p. 80).

problem which might lead to the idea that the determining act of thought, the 'I think', cannot be identified with the subject, Kant claims that we merely represent this spontaneity and ourselves as "self-active beings" and we are justified in doing so as this act of determination by thought is the reason that a subject can refer to itself as 'intelligence'.[21]

While agreeing with Kant that the highest act which is thought, the determination of the subject's existence, is realized in time Deleuze rejects Kant's too easy attribution of this act to the subject. If it is impossible, as Kant admits, to determine the act of determination itself, then one cannot know that it originates from a spontaneous subjectivity.[22] But accepting that the act of thought cannot be located in time, that it is not one event among others does not mean that it bears no relation to time. The act of determination that can be realized only in time involves also the form of time. Thought, then, is not within time but rather generates the form of time. This notion of thought shows that like time, thought does not originate from the subject and yet this insight is reached through delving into subjectivity.

Now, we can order the syntheses of time as advancing from the contraction of instants in the living present, to the comprehension of all the presents, of all events, within a pure past. Repetition's act in the third synthesis applies, then, to time itself – it is the form of time. What would be the form of time? this can be answered from a perspective where time is "liberated from its overly simple circular figure, freed from the events which made up its content, its relation to movement [is] overturned; in short, time is presenting itself as an empty and pure form. Time itself unfolds… instead of something unfolding within it".[23] This pure and empty form is, as we saw above, a determining act, and what it determines here is a distinction between a 'before' and an 'after' setting by that the inner form of time. This determination of a before/after distinction is what Deleuze calls the form of change, and thus the form of time is the form of change; indeed Deleuze claims that "Time is the form of change, but the form of change does not change".[24]

This third synthesis should be thought of as a synthesis of the future for several reasons one being that it is aimed at systematically warrant the openness of experience to the new. From the perspective of the second synthesis of the pure past, experience is closed upon itself containing already everything that can happen. Thus we can say that for Deleuze it is not the case that time is open to the future but rather it is the future that opens time. This synthesis, then, can be seen as a metatemporality operating from within time, synthesizing it as dynamic in a way that

---

[21] The text Deleuze refers to is the note in section 25 of the transcendental deduction of Kant's *Critique of Pure Reason*. See Kant (1998, B157–158) and Deleuze (1994, p. 86).

[22] Deleuze argues in this context that subjectivity is split not between the active and the passive as it is usually understood but within passivity because the auto-affection of the 'I think' occurs to a passive subject who cannot be justified in identifying herself with this 'I' or act of thought. Thus, the passive subject is internally fractured. See Deleuze (1994, p. 86).

[23] Deleuze (1994, p. 88).

[24] Deleuze (1994, p. 89). For a recent discussion of the heterogeneous sources of Deleuze's third synthesis, see Voss (2013).

cannot be found in the first and the second syntheses. These two first syntheses manifest the rhythm of entities and the movement of present events awaiting actuality within a past in which they are already contained while the third does not apply to the content of time but to time itself.

If we look, then, at the three syntheses we get a complex and rich notion of time that provides a pulse or a rhythm in the first synthesis, a passage and movement together with containment and continuity in the second and a change as a determination of a before/after distinction in the third which opens time altogether for the new but also characterizes time through its form as generating change.

12.3 Deleuze's Time and McTaggart's Series

I would like to conclude with Deleuze's relation to the transcendental tradition and then to deepen our understanding of Deleuze's notion of time with the help of the McTaggartian perspective on time.

We saw that Deleuze should be read as thinking about time within the transcendental tradition: Deleuze accepts that the question of time should be dealt with in the context of a theory of experience, and that time cannot be derived from experience but is presupposed by experience. He also agrees that a first move would be to turn to subjectivity in order to investigate the nature of time and hence he apparently accepts the transcendental hierarchy between the subjective and the objective notion of time.

But this turn to subjectivity brings Deleuze to argue, first, that the subject itself is passive, hence the syntheses of time do not originate from the subject but provides the framework within which a certain type of subjectivity is formed, and second that we should reject the identification of ideality with subjectivity, or rather the exclusivity of the subject with regard to ideality. Thus instead of challenging the transcendental attempt to justify a hierarchy between the subjective and the objective notions of time, Deleuze challenges the distinction altogether.

Let us think, for example, on the first synthesis of time – the living present: it does not manifest any unique human or what we would traditionally call subjective traits. The repetitive nature of the act of contraction is such that it brings about a change within a contemplating mind on the organic level. This repetition is, according to Deleuze, ideal, an ideality that subsists within the organic level (which Deleuze refers to as 'contemplation'). The entity within which contemplation occurs is made different through having a rhythm.[25] Is this living present internal?

[25] Deleuze's use of the term 'contemplation' is unique and is not based on the traditional sense of contemplation as related to the supreme intellectual capability of focusing one's mind on intelligible beings or ideas. Rather, contemplation refers to an ideal aspect that should be attributed to entities already on the organic level as passively occurring in them. To contemplate is to draw something from something else, and this involves first, ideality and second externality. That is, a contemplating mind can be such only if there is another ideality from which a difference is drawn, or better, by which a difference is made in the contemplating mind. This is what happens in the

Is it external? This is exactly the point: it must be thought of as ideal but it demands an irregular understanding of ideality, an ideality that is not grafted upon subjectivity (as would be, for example, a Hegelian type of such an ideality) and does not set a limit to all possible experiences or provides an anchor to possible meanings within a world; rather it should be thought of as productive of these experiences and meanings, including the subjectivity that is capable of having these experiences and entertain those meanings.

Thus we can see that although Deleuze holds that time is ideal, we cannot simply attribute to him a position like "transcendental idealism", that is, the thesis that time (together with space) is the subject's form of intuition and even in a broader sense, according to which the subject is thought of as the origin of time. On the other hand, Deleuze does not claim that time should be thought of as a thing in itself but rather that not being a thing in itself does not make time dependent upon the subject. Time, then, is liberated from subjectivity (but not detached from it) and the same goes for the relation of ideality to subjectivity. This means that in Deleuze's case time's being ideal does not prevent it from directly apply itself to processes in the world as it is not mediated through subjectivity and projected as a kind of an interpretative scheme upon the subject's surrounding environment, be it human history or cosmological history. Time is not merely the negative condition of all possible phenomena but rather becomes, from a Deleuzian perspective, a generating internal condition for all phenomena, producing phenomena differentially according to the relevant level of synthesis. This applies also to time itself: the future – change as the form of time – injects change to what is otherwise a closed and predictable system of experience, hence it operates as the productive principle of time itself.

A last point regarding Deleuze's relation to the transcendental tradition is that due to the synthetic structure of Deleuzian time it makes no sense to determine whether time is finite or infinite: the living present is organically bound, hence it is finite, but the second synthesis is placed within a transcendental memory that can be thought of as infinite or embracing the whole of time, while in the third synthesis there is no sense at all to think of time as finite or infinite as this synthesis is of time's nature as the form of change. The whole of this structure generates different temporal levels that fit different layers of experience which correspond, as we saw above, to different faculties. This structure allows for different and even competing conceptions of time (that emerge, I believe, from the characteristics of the second synthesis both the active and the passive). By this Deleuze also retains a transcendental trait of the discussion of time which is committed to provide a structural reason for the possibility of thinking, intuitively, as it where, about time as objective in the regular sense of the term. But most importantly it allows for a differential

---

relation between repetition and the rhythm that generates the organic contemplating mind. Deleuze refers this position where "all is contemplation" to Plotinus, see Deleuze (1994, p. 75). This, however, should not be taken as meaning that Deleuze is committed to the background metaphysics of Plotinus. It is more likely that this specific use suits his more general project of establishing relations between different levels of ideality that are not attributable to subjectivity.

notion of time, a differentiation that is inseparable from the nature of repetition on each level of synthesis.

Now, if we turn back to the question, whether there would be time had there been no subject, I believe that Deleuze's answer, in contrast to his transcendental predecessors, would be definitely positive as it follows from the distinction between subjectivity and ideality. This strong assertion should be qualified, however, because as we saw, at least in the first synthesis, human subjectivity does not exhaust all forms of subjectivity. Subjectivity itself is differential and layered due to the different syntheses of time; hence, the above can be reformulated as follows: time would still be even if there would be no self-conscious subject, or in other words, time is not dependent upon that kind of subjectivity. In light of Deleuze's notion of time the question should be modified and we should ask whether there would be time without a world and I think that with regard to this question the answer would be negative. Yet, this answer is also grounded in transcendental reasoning: the whole activity of time or of repetition through the syntheses of time is the production of experience; thinking of time, then, without a world would imply a sterile, idle time which is impossible if we take time together with Deleuze as directly operating within experience. A different question would be whether subjectivity itself in its active sense, having been dethroned from its status as the origin of time, should be thought of at least as necessarily deriving from this notion of time. Here, I believe that a probable Deleuzian answer would be that given the productive character of time some kind of subjectivity might ultimately emerge, because as we saw subjectivity in the sense of an active subject is merely a certain modification of a more comprehensive ideality.

Finally, I would like to shed light on the idea of "change's order" or the form of change in Deleuze's thought of time. In order to do that, I will focus on the third synthesis with the help of McTaggart's famous argument for the unreality of time.[26]

According to McTaggart's argument, time consists of two ways of determining the positions of events: The first distinguishes between past, present and future (the A series) while the second distinguishes between earlier and later (the B series). McTaggart then argues that it is the A series that consists of change and assuming that time implies change, the A series is essential for the reality of time.[27] In other words, the reality of time hinges upon the possibility of defending the past-present-future distinction as logically valid. Following this line of argument, the B series is the result of the combination of the A series, which consists of change, with another kind of series, the C series, which is not temporal "for it involves no change, but

---

[26] McTaggart (1908). McTaggart's position regarding the unreality of time is rarely accepted but the two series (A and B) which are argued to be constitutive of time became since the basis for two rival theories regarding the nature of time. See Dyke (2002) and Oaklander (2004, pp. 17–36, 51–62) for discussions of McTaggart's argument and the debates between A- and B-theorists. This is not important for the purpose of our present discussion.

[27] In McTaggart's own words the A and B series are both essential to time yet not equally fundamental: only the A series is ultimate (as well as the a-temporal C series that is discussed in what follows), see McTaggart (1908, p. 463).

only an order".[28] An example of a C series is the order of the letters in the alphabet: this order is indifferent to the direction of reading – it can be read either from A to Z or from Z to A while the order of the letters, their relative position, does not change. McTaggart even argues that the earlier-later relation varies with the direction of reading: while A is earlier than B in one reading direction, it is later than B if the letters are read in the opposite direction. Change, argues McTaggart, lies in the A series, and it is the reality of change with regard to events that he presents as contradictory. McTaggart's argument with regard to the unreality of time, however, is not my concern here, but rather his understanding of change.

First, it is clear, that the sense of 'past', 'present' and 'future' viewed from a transcendental perspective cannot be given the same sense as in McTaggart: in the first synthesis, past and future are but aspects of the present and not times separated from it. In the second synthesis it is impossible to genuinely speak of the future. The way in which the past is synthesized – through transcendental memory –does not leave room for a future. In the third synthesis the future, understood as the form of time, applies itself to the closed past such that one cannot speak of simple continuous relations between past, present and future, thus no passage between A-properties in McTaggart's sense is formed. From this it follows, that from the transcendental perspective argued for by Deleuze, the A series are denied not in favor of a notion of time based on the B series (as some kind of B-theory would suggest) but because an A-theory requires relations between past, present and future that can be established only in the active second synthesis which has no transcendental import and conceals rather than reveals the nature of time.

However, McTaggart claims that "…where there is no A series there is no time".[29] Thus, it would help us clarify Deleuze's notion of time if we see whether the fact that no A series in McTaggart's sense can be established from this transcendental perspective is problematic or rather reflects a disagreement over the background assumptions that support McTaggart's claim. I believe it is the latter and I think that this disagreement allows us to see the full import of Deleuze's sense of temporal change and its contribution to the understanding of time.

McTaggart's notion of time can be characterized as external and as passing. By characterizing time as 'external' I refer to McTaggart's claim that "changes must happen to the events of such a nature that the occurrence of these changes does not hinder the events from being events, and the same events, both before and after the change".[30] Change, that is, time with its A-properties of past, present and future should not interfere with the identity of an event. For McTaggart, then, events are *fait accompli* such that if A-properties were to be added to them it would not change anything about them except for the change itself, that is, the A-property: the same event has been future, became present and turned finally into past. Yet, this means that changes are external to events and do not really change anything within them. From a Deleuzian perspective this would mean that time is sterile – what exists,

[28] McTaggart (1908, p. 462).

[29] McTaggart (1908, p. 461).

[30] McTaggart (1908, p. 460).

exists in such a way that it can be accounted for independently of time. From these assumptions McTaggart had drawn the justified conclusion that it just might be impossible to add time to existence.[31] Under such assumptions time could be attached to existence only if time would be understood as a medium of transportation that makes no internal difference but only moves events or things from one place to another along the timeline. This movement that does not change anything is exactly what McTaggart calls "change". Thus, McTaggart actually reduces change to movement and at the same time exhibits the limited nature of movement. This means that the essential character of time is its passing; if time does anything at all – it is to pass.

Now we can clarify Deleuzian time by pointing to the disagreement. First, we saw that the externality of time means necessarily that time is sterile, and we saw on the other hand that Deleuze's notion of time is based on the understanding of time as productive with regard to experience. Second, the idea that events are completed in themselves independently of some temporal characterization is widely accepted in the philosophical tradition. This is due to the logical or conceptual perspective taken with regard to events. This seems, however, odd: if there is something that distinguishes between events and things, it is that events are temporal to begin with. How the temporality of events should be understood might be open to debate, but what Deleuze has in mind is that the change as an act of determining a temporal difference of 'before' and 'after' is the form of time. That which makes events what they are is derived from the form of time, in other words, time itself has the form that is manifested in events. This means that time cannot be thought of as external to events and events cannot be thought as being independent of temporality. Actually, Deleuze just might have the opposite problem, that is, while McTaggart thinks of events in terms of things, Deleuze extends his dynamic temporal thought over things, such that they are ultimately viewed as events.[32] Thus, while the earlier-later distinction only describes externally the given order of events, the before-after distinction orders time from within thus affecting also the events that are ordered in time.

This brings us to the last point which is the character of time as passing, or what is called "the passage of time". As you can see by now, Deleuze rejects the reduction of change to movement and accordingly he rejects the passage of time as the most fundamental feature of the temporal. Passage as understood above merely transports

[31] Indeed, such a view led Merleau-Ponty to the conclusion that from this kind of perspective "the very notion of event has no place in the objective world…if I consider the world itself there is simply one indivisible and changeless being in it" (Merleau-Ponty 2002, p. 477). Perceiving events as reified hence, independent of time with McTaggart or claiming that in such case there is no use of speaking of events at all like Merleau-Ponty, amounts to the same.

[32] This view is mostly developed in Deleuze's *Logic of sense* (Deleuze 1990) where the notion of event becomes central and goes hand in hand with a notion of 'becoming' (*devenir*). It should be noted, however, that Deleuze's notion of event in the *Logic of Sense* is not the empirical notion of event from which time is liberated in the third synthesis of time in *Difference and Repetition*. This shows that the conceptions of time elaborated in these works should not be automatically treated as one and the same.

a thing or an event from one place to another, leaving it intact. But change acts from within as a distinction of a 'before' and 'after' which generates an inner dynamic. Think about everyday photos that you can see in the newspaper of a 'before' together with an 'after': there is little sense in thinking of the difference between the photos on the model of a movement from one place to another. What operates behind the scene, literally in this case, is an act that makes a difference, which is the sense of change for which Deleuze argues. Although for Deleuze the third synthesis of time applies to time itself being the form of time, this example might illustrate the sense of this temporal change as making a difference of 'before' and 'after'. And being a philosopher of difference it should come to us as no surprise that Deleuze's notion of time as the order of change is one of making a difference.

References

Couzens Hoy, David. 2009. *The time of our lives: A critical history of temporality*. Cambridge, MA: The MIT Press.

De Warren, Nicolas. 2009. *Husserl and the promise of time: Subjectivity in transcendental phenomenology*. Cambridge: Cambridge University Press.

Deleuze, Gilles. 1990. *The Logic of Sense*. Trans. Mark Lester and Charles Stivale. New York: Columbia University Press.

Deleuze, Gilles. 1991a. *Bergsonism*. Trans. Hugh Tomlinson and Barbara Habberjam. New York: Zone Books.

Deleuze, Gilles. 1991b. *Empiricism and Subjectivity: An Essay on Hume's Theory of Human Nature*. Trans. Constantin V. Boundas. New York: Columbia University Press.

Deleuze, Gilles. 1994. *Difference and Repetition*. Trans. Paul Patton. New York: Columbia University Press.

Dyke, Heather. 2002. McTaggart and the truth about time. In *Time, reality and experience*, ed. Craig Callender, 137–152. Cambridge: Cambridge University Press.

Grosz, Elizabeth. 2013. Habit today: Ravaisson, Bergson, Deleuze and us. *Body & Society* 19: 217–239.

Heidegger, Martin. 1962. *Being and Time*. Trans. John Macquarrie and Edward Robinson. Oxford: Blackwell.

Husserl, Edmund. 1991. *On the Phenomenology of the Consciousness of Internal Time (1893–1917)*. Trans. John Barnett Brough. Dordrecht: Kluwer Academic Publishers.

Kant, Immanuel. 1998. *Critique of Pure Reason*. Trans. Paul Guyer and Allen W. Wood. Cambridge: Cambridge University Press.

Lord, Beth. 2012. Deleuze and Kant. In *The Cambridge companion to Deleuze*, ed. Daniel W. Smith and Henry Somers-Hall, 82–102. Cambridge: Cambridge University Press.

McTaggart, John Ellis. 1908. The unreality of time. *Mind* 17: 457–474.

Merleau-Ponty, Maurice. 2002. *The Phenomenology of Perception*. Trans. Collin Smith. London: Routledge.

Michalski, Krzysztof. 1997. *Logic and Time: An Essay on Husserl's Theory of Meaning*. Trans. Adam Czerniawski and James Dodd. Dordrecht: Kluwer Academic Publishers.

Meillassoux, Quentin. 2008. *After Finitude: Essay on the Necessity of Contingency*, trans. Ray Brassier, London: Continuum.

Oaklander, L. Nathan. 2004. *The ontology of time*. New York: Prometheus Books.

Turetzky, Philip. 1998. *Time*. London: Routledge.

Voss, Daniela. 2013. Deleuze's third synthesis of time. *Deleuze Studies* 7: 194–216.

Williams, James. 2011. *Gilles Deleuze's philosophy of time: A critical introduction and guide*. Edinburgh: Edinburgh University Press.

MIX
Papier aus verantwortungsvollen Quellen
Paper from responsible sources
FSC® C105338

If you have any concerns about our products,
you can contact us on
ProductSafety@springernature.com

In case Publisher is established outside the EU,
the EU authorized representative is:
Springer Nature Customer Service Center GmbH
Europaplatz 3, 69115 Heidelberg, Germany

Printed by Libri Plureos GmbH
in Hamburg, Germany